WHAT DOES
IT MEAN TO BE
HUMAN?

WHAT DOES IT MEAN TO BE HUMAN?

RICHARD POTTS AND **CHRISTOPHER SLOAN**

NATIONAL GEOGRAPHIC

WASHINGTON, D.C.

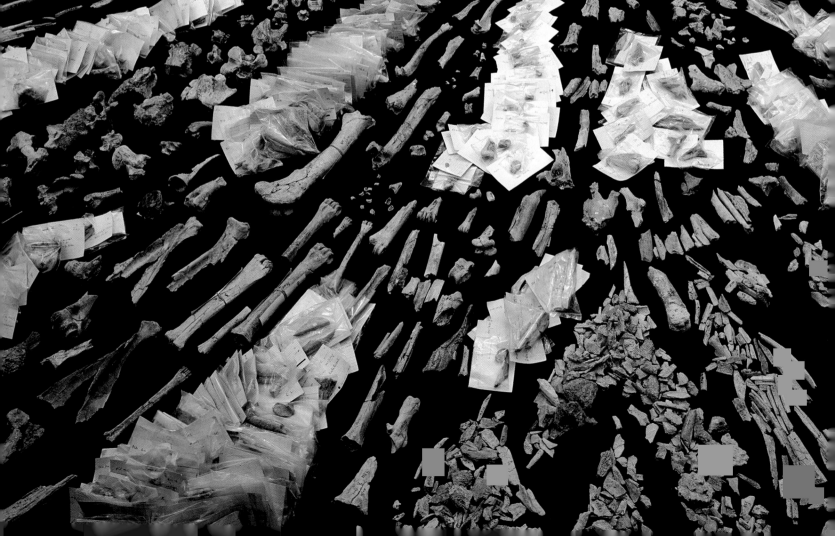

CONTENTS

Page 2: *A meeting of human and chimpanzee hands reflects the discovery of our connection with the natural world.*

Opposite: *Stone tools by the thousands (top), found among broken antelope bones at Olduvai Gorge in Tanzania, provide evidence of an early innovation in our lineage.*

SMITHSONIAN ACKNOWLEDGMENTS

The David H. Koch Hall of Human Origins and the Human Origins Initiative of the Smithsonian Institution's National Museum of Natural History are made possible through the generous support of the following individuals and organizations:

David H. Koch

Peter Buck

The National Science Foundation

The Ruth and Vernon Taylor Foundation, Montana

Mr. and Mrs. Jeffery Meyer

Charles Dern

Joan and Robert Donner

|||||||||||||||||||||||||||||||||||||||||||||||||||||||||||||||||||

The following individuals from institutions outside of the Smithsonian have generously contributed time and expertise in reviewing content and providing objects featured in the David H. Koch Hall of Human Origins:

Ernesto Abbate, Leslie Aiello, Takeru Akazawa, Zeresenay Alemseged, Nira Alperson-Afil, Susan Antón, Juan Luis Arsuaga, Fachroel Aziz, Elizabeth Babcock, David Begun, Connie Bertka, C. K. Brain, David Braun, Peter Brown, Raymonde Bonnefille, Joseph F. Chance, Deng Chenglong, Kevin Cole, Chris Collins, Nicholas Conard, Francesco d'Errico, Darryl de Ruiter, Jussi Eronen, Dean Falk, I. O. Farah, Xie Fei, Mikael Fortelius, Diane France, Alain Froment, Naama Goren-Inbar, Fred Grine, William Harcourt-Smith, Katerina Harvati, Christopher Henshilwood, Andrew Hill, John F. Hoffecker, Ralph Holloway, Huang Weiwen, Jean-Jacques Hublin, Michèle Julien, William Jungers, Petro Keene, André Keyser, William Kimbel, Kathelijne Koops, Ravi Korisettar, George D. Koufos, Benson Kyongo, Meave Leakey, Dennis Liu, Liu Wu, David Lordkipanidze, Blaine Maley, Alan Mann, Giorgio Manzi, Tetsuro Matsuzawa, Louise Mead, Emma Mbua, William C. McGrew, Jeff Meldrum, James B. Miller, Anne Nivart, Thomas Plummer, Jill Pruetz, Kaye Reed, Hélène Roche, Claudia Rodrigues-Carvalho, Lorenzo Rook, Christopher Ruff, Sileshi Semaw, John Shea, Kathy Schick, Chris Sloan, Tanya Smith, Chris Stringer, R. Sukhyar, Thomas Sutikna, Ian Tattersall, Ana Gracia Téllez, Francis Thackeray, Anna Thanukos, Erik Trinkaus, Peter Ungar, Marian Vanhaeren, Mamitu Yilma, Alan Walker, Carol Ward, Kurt Wehrberger, Wang Wei, A. Djumarna Wirakusumah, Sibylle Wolf, Sarah Wurz, Zhu Rixiang.

Cristián Samper

One of our most fundamental questions relates to our own origins. This book gives us some answers. *What Does It Mean to Be Human?* takes us on the epic journey of the human species and explains how the characteristics that define human beings today evolved over the past six million years. It explores how our unique physical traits, behaviors, and ways of interacting with our surroundings emerged as our ancestors struggled to survive in a dramatically changing world.

This book is published at a time of rapid-paced discoveries that afford scientists an unprecedented opportunity to piece together a new understanding of human origins. Rick Potts, the Peter Buck Chair in Human Origins at the Smithsonian Institution's National Museum of Natural History, collaborated with *National Geographic* magazine paleoanthropology editor Chris Sloan to tell this story.

A few years ago, I had the opportunity to visit the Smithsonian's research site at Olorgesailie (Kenya) and see it with Rick Potts's help. Hiking for days in that arid landscape, my wife and I glimpsed the million-year-old world of our resourceful ancestors, who created the array of tools that first helped early humans master their environment. This book offers you a look into that world, which not many people are as privileged to see as I was.

What Does It Mean to Be Human? is also designed to serve as a companion to the David H. Koch Hall of Human Origins at the Smithsonian's National Museum of Natural History, which opened on March 17, 2010, on the 100th anniversary of the museum. Neither the exhibition nor this book would be possible without the generous support of many generous people, especially David H. Koch and Peter Buck. Their commitment to science, education, and our understanding of human origins is an example to us all.

I trust this book will help us understand our human journey over the past six million years in response to a changing world, and I invite you to come and visit the David H. Koch Hall of Human Origins at the Smithsonian in the near future.

—Cristián Samper
Director, National Museum of Natural History

THE MEANING OF AN EVOLUTIONARY ORIGIN

WHAT DOES IT MEAN TO BE HUMAN? SPEAKING TO AUDIENCES ACROSS THE COUNTRY about human evolution, Rick Potts often asks people to reflect on what this universal question means to them. Their answers are diverse— and sometimes surprising: We create and appreciate beauty. . . . To damage the world around us and have the choice to care. . . . I've got the ability to operate a remote control. . . . Not much hair over our bodies—except some people I know. . . . Knowing that God loves me. . . . To have thumbs, wonderful thumbs! . . . We can remember our ancestors and think of the future. . . . Every one of these statements reminds us that we all have thoughts and feelings about who we are and where we came from. At the same time, scientists are piecing together evidence that throws new light on the origin of our human species.

Opposite: *This detail from Gustav Klimt's "Tree of Life" (1905–09) depicts the artist's fanciful interpretation of the universal symbol.*

The question of origins touches the deepest roots of human curiosity. For thousands of years, the diverse creeds and sagas of societies everywhere have shaped people's understanding of their origins and defined their identity as humans. The discovery of new evidence concerning our origins thus poses a challenge. A science that can alter how we think about our own origins, and about ourselves, has the potential to be embraced with excitement . . . or ignored and even ridiculed for the audacity to brush up against core beliefs about our place in the world.

Although others had previously thought of the idea of evolution, it was Charles Darwin's synthesis of evidence on the origin of species that captured both the public's excitement and its resistance. Darwin applied his carefully recorded observations and experiments to the hypothesis that new species arise over time through natural processes. He struggled personally with the opposition of his findings to the belief that all species were created by God at one time. Darwin proposed that populations become adapted to their surroundings over time, a process he called natural selection, and that this process is responsible for the formation of new species and thus for all of biological diversity. The "big idea" of evolution is that all organisms are related to one another, having

This "tree of life" drawing from one of Charles Darwin's 1837 notebooks reveals a great mind refining a great idea: Living plants and animals represent descent with modification from a common ancestor.

shared a sequence of common ancestors back through time. Life is an enormous tree of kinship among organisms rather than an array of individually created forms. Darwin also realized that evolution involves extinction: Certain ways of life ultimately meet their demise. Thus, the branching tree of species has been pruned over the eras of life on Earth.

When Darwin died in 1882, the evidence available to support his hypothesis was sparse. One difficulty was the poor fossil record. Transitional forms between species, especially major groups of organisms, were very rare. More than three decades after the publication of *On the Origin of Species,* no one really knew if Earth's history was long enough for evolution to have taken place. The foremost physicist of the time, Lord Kelvin, claimed that our planet's age could be no more than a few tens of millions of years old, based on the rate of Earth's cooling from a molten state to its present temperature. This estimate could have been fatal to Darwin's hypothesis, as he suggested that hundreds of millions of years might be necessary for Earth's diversity of species to evolve by gradual processes of change.

In 1882, no one knew of DNA, the molecule that bears the genetic code within living organisms. Nor did anyone know that DNA would offer a powerful tool for testing the relationships among all living things. Experimenting with garden peas, the Augustinian monk Gregor Mendel had already described, in a largely unread publication titled "Experiments With Plant Hybrids," the basic laws of inheritance. But neither Darwin nor anyone else was aware of the existence of genes or of how their chemical makeup could continually build a storehouse of variations—a gene pool—from one generation to the next. All Darwin realized was that novelty and variation, what we now understand as mutations and new combinations of genes, were essential to his hypothesis. Some means had to exist, if his idea was correct, for creating the immense diversity of raw material as a starting point for shaping the characteristics of organisms by natural selection and, eventually, the enormous variety of species. Although Darwin experimented and described novel variations, with some better than others at helping organisms survive and reproduce, no one had yet made field observations that demonstrated how important natural selection is in the wild.

Any number of uncertainties could have demolished his hypothesis. His contributions to biology are celebrated, however, because findings across the sciences, from physics and chemistry to paleontology and genomics, offer overwhelming evidence of the processes of adaptation to the environment, the relatedness of all organisms, the origin and extinction of species over time, and the subsequent rises and falls in biological diversity. Science is a process of critical revision, and the past century of tests of Darwin's hypothesis has largely magnified his original vision.

The fossil record of human evolution was essentially nonexistent at the time of Darwin's death, with only a handful of Neanderthal fossils known. Today, based on fossils that range from isolated teeth to nearly complete skeletons, approximately 6,000 fossil individuals of early humans, spanning the past six million years, are known. For any group of large mammals, this is a good record of fossilized bones, and it enables students of the field to trace many critical transitions in brain and body size, skull shape, and changes in the details of the teeth and the skeleton. Advances in medical technologies such as CT (computed tomography) and SEM (scanning electron microscopy) now allow close examination of the internal and surface details of fossils. These developments yield insights, for instance, into how fast ancient human infants grew up and how different diets etched distinct pits and grooves on early ancestors' teeth.

Humans share genes with all living organisms thanks to billions of years of evolution. The percentage of genes, or DNA, that organisms share shows their similarities and how close their relationship is in the tree of life.

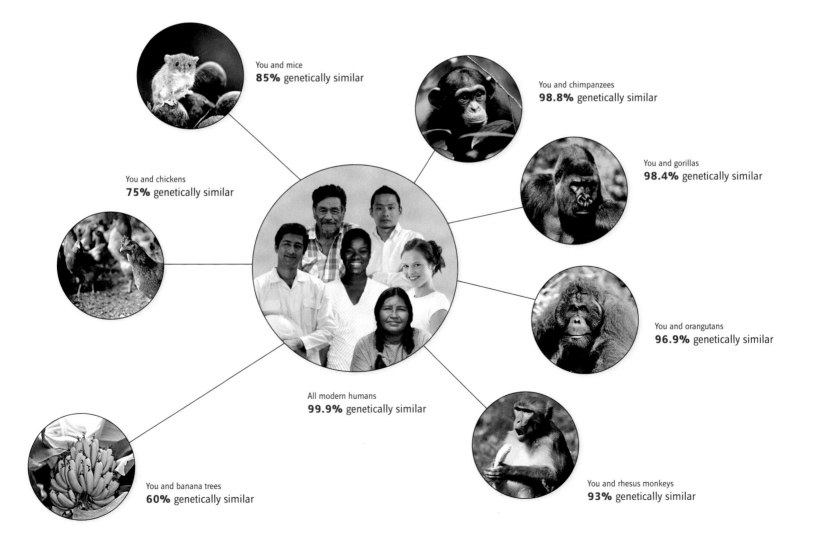

You and mice
85% genetically similar

You and chimpanzees
98.8% genetically similar

You and chickens
75% genetically similar

You and gorillas
98.4% genetically similar

You and orangutans
96.9% genetically similar

All modern humans
99.9% genetically similar

You and banana trees
60% genetically similar

You and rhesus monkeys
93% genetically similar

Origins around the world

Universal curiosity about human origins is manifested in sacred stories from around the world. The following examples only begin to span the spectrum of beliefs on how humans came to be and on our place in the cosmos.

Kayan, northwestern Borneo: A vine from the moon marries a great tree that sprouted from wood dropped from the sun; the tree gives birth to twins who marry and become the parents of the people on Earth.

Yoruba, Nigeria: Among the gods of the sky, Obatala creates the land, makes people out of clay, and bakes them in the sun. The supreme Olorun later offers people the breath of life, enabling them to do the things humans do.

Quiché Maya, Guatemala: After trying unsuccessfully to make humans out of mud and then wood, the creators succeed with maize (corn).

Persia (Iran): The first human couple emerged from a reed that grew from the blood of a slain primal man.

In addition, thousands of archeological sites, defined by clusters of tools and other artifacts too numerous to count, bear the evidence of how human predecessors made their technologies; where they roamed to get stone for making tools; how they procured certain foods; and when they began to control fire. Stone tools provide robust clues about the geography of human origins and the environments in which our ancestors lived. Sculpted figurines, rock paintings, and shell necklaces also left behind by our ancestors show when symbolic behavior became a critical part of their lives and when communities began to differentiate and interact.

As scientists have worked out the physical and chemical principles of geological time, Darwin's suggestion that the evolution of life's diversity required several hundred million years has been supplanted by evidence that life evolved over four billion years. For the period of human origins, more than a dozen techniques for dating rocks that bear the fossil evidence, or the fossils themselves, can now be applied and cross-checked. The dating of human origins has become more refined than scientists a century ago could have imagined.

It was not until the 1950s that the structure of DNA and how it functioned were discovered, paving the way for tests that determine the genetic relationships among individuals or species. Darwin knew nothing of genetics. But, using the principles of evolution, he predicted that the common ancestor of human beings and other primates would be found in Africa, where the African apes live. It might have turned out that these apes would possess DNA most similar to a lemur's or another distantly related primate's; instead, among all the species of mammals, chimpanzee and gorilla DNA are most akin to our own.

After many decades of fossil hunters searching the continents, we know now that the oldest human fossils come exclusively from Africa, which is often called the cradle of humanity. As new finds bring us closer to the six- to eight-million-year-old divide when, based on DNA studies, the chimpanzee and human evolutionary branches diverged, fossil bones on the human side of the divide appear increasingly apelike. While paleontologists continue to find the fossils that bridge from land mammals to whales, from fish to higher vertebrates, from the "terrible lizards" of the dinosaur world to birds, researchers are also filling in more and more of the key transitions in human evolution.

EVOLUTIONARY MILESTONES

The hunt for fossil treasures continues at an amazing pace, and researchers have added at least a half dozen new fossil lineages to the human family tree over the past two decades. But you will not find in this book a standard account of the search for fossils and the excitement of stunning finds. The theme of this book, instead, concerns the milestones or benchmarks in the lengthy process of becoming human—in other words, the emergence of the characteristics that define the shared evolutionary history of all human beings. We can now say with certainty that walking upright evolved before our ancestors became dependent

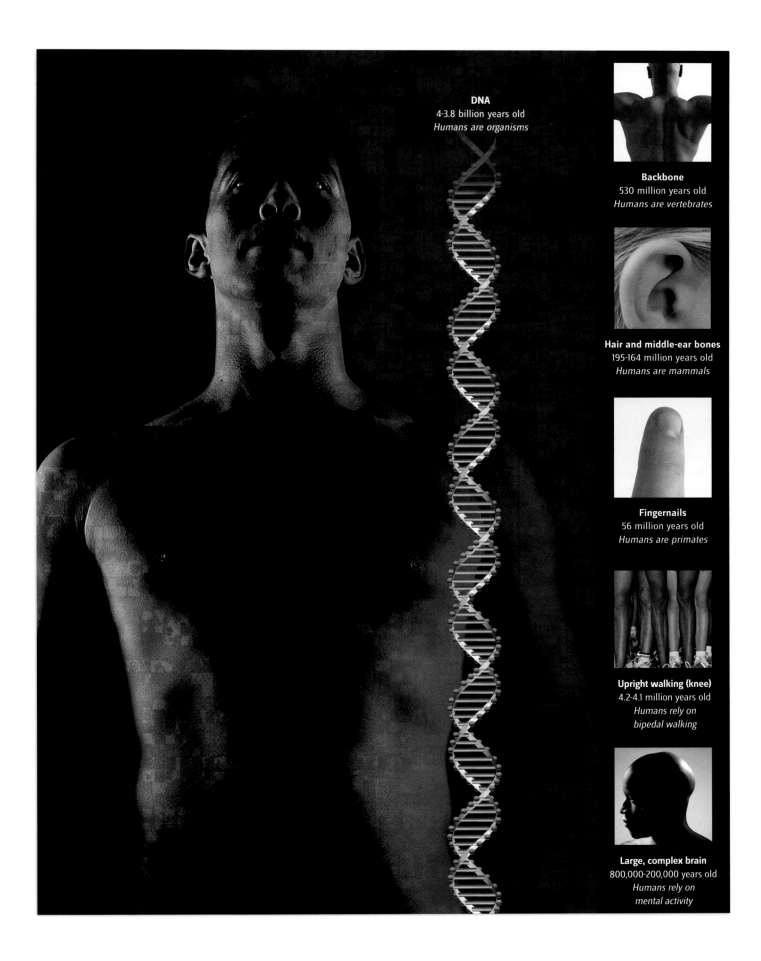

DNA
4-3.8 billion years old
Humans are organisms

Backbone
530 million years old
Humans are vertebrates

Hair and middle-ear bones
195-164 million years old
Humans are mammals

Fingernails
56 million years old
Humans are primates

Upright walking (knee)
4.2-4.1 million years old
*Humans rely on
bipedal walking*

Large, complex brain
800,000-200,000 years old
*Humans rely on
mental activity*

on tools. And both of these developments took place long before the evolution of a very large brain and, later still, the capacity to produce symbolic artifacts and art. These defining features of our kind dawned in small groups of early humans who foraged each day for food and water. Only later did the transition to an agricultural way of life take place, which enabled food to be stored and the people who sowed and harvested it to feed others. Now home to billions of people, the world as we know it is based on this lengthy series of milestones, shaped by survival and change in a series of ancestors over millions of years in response to a changing world.

The standard telling of the story of human origins as one of ascent and success has taken a sharp turn.

Responding to environmental change has been a central feature of our ancestors' survival, we now know. The six million years of human evolution comprise one of the most dramatic eras of climate instability in Earth's history. This finding is surprising to those who think that the ancient habitat of our ancestors has already been narrowed down to a dry African grassland or a frigid European ice age. In fact, one of the hallmarks of our evolutionary story is that the conditions of survival kept changing. In every generation, groups of early human foragers faced the problem of how best to endure in their immediate surroundings. But over time, the line between thriving and extinction was a matter of how best to adjust to the uncertainties of climate shifts between rainy and arid, cold and warm. The coupling of these two exciting domains of science—past climate change and the origin of humans—provides a new and compelling way of examining the evidence of human evolution.

The standard telling of the story of human origins as one of ascent and success has taken a sharp turn. With many new fossils and lineages of early human species added to our genealogy, the fact remains that only one human species has survived, and the diverse ways of life that characterized earlier ancestors and our evolutionary cousins are now extinct. This is not simply a curious fact of passing interest. Genetic studies point to the near demise of our own species—at a time in the past 100,000 years when the breeding population of *Homo sapiens* declined to only a few thousand adults. The fragility of human life, when viewed in its evolutionary setting, is one of the implications of recent research into human origins. This perspective is also likely to surprise anyone who thinks of our kind as endowed with dominion over nature.

CORE CONCEPTS OF HUMAN EVOLUTION

The study of human evolution, or paleoanthropology, is poised at the intersection of several scientific fields. Finding out the age of sites depends on essential knowledge in physics and chemistry. Understanding the order of evolutionary events over time rests on the foundations of geology. Bringing environmental records to bear on human evolution, and figuring out the changing settings of natural selection, rely on diverse research approaches in the Earth sciences. The study of fossils, the identification of ancient species, and their arrangement in a family tree all call upon the basics of biology. The intimate relatedness of

Homo sapiens and earlier species of humans to the African apes, and to all other primates through a series of common ancestors back through time, demands knowledge of the founding principles of genetics and the study of DNA. The fact of human evolution is based on a foundation of knowledge spanning the array of scientific endeavors.

Thus the grave doubts and denials sometimes expressed in the public arena over whether humans evolved, or about whether children should be exposed to this well-established body of knowledge, are painfully hard for students of human evolution, and scientists in general, to understand or accept. From their perspective, such denial seems to undermine the overall goal of science education and devalue what is gained by discovering new things about the world.

The challenges posed by evolution

In the vibrant scientific field of human evolution, new discoveries and research findings are regularly reported as lead stories in newspapers and other media. Despite strong public interest, however, many people find the idea of human evolution troubling from a religious perspective. While polarized public opinion on the matter is highlighted, the diversity of contemporary religious responses to evolution is less recognized. These responses point to opportunities for a productive relationship between science and religion without assuming a conflict between the scientific evidence of human evolution and religious beliefs.

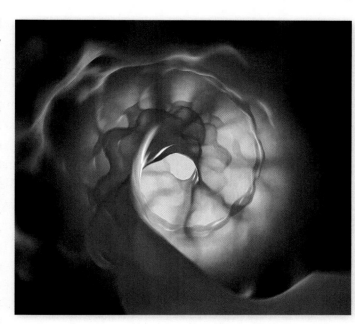

DNA's double helix, shown here in an artist's concept, is a source of ongoing scientific discoveries concerning evolution and the history of life.

strongest conflicts develop when either science or religion asserts a standard of truth to which the other must adhere or otherwise be dismissed. An alternative approach sees engagement between science and religion as positive. Engagement takes many forms, including personal efforts by individuals to integrate scientific and religious understandings, statements by religious organizations that affirm and even celebrate scientific findings, and constructive interactions between theologians and scientists seeking common ground, respect, and shared insight into how the science of human evolution contributes to an awareness of what it means to be human.

There are a number of different approaches to the science-religion relationship. One approach is to see science and religion as separate domains that ask different questions—for example, about the natural world in science and about God in religion. This approach depends on respecting and maintaining the distinctions but can sometimes overlook the ways in which scientific interpretations may have an effect on religious beliefs. Conflict can arise when efforts are made to eliminate the separation that the first approach assumes. The

Surveys on the public acceptance of evolution indicate that conflict continues to impede public understanding of scientific methods and ongoing discoveries. Looking beyond that, however, the wider variety of perspectives suggests that there is considerable support for maintaining the integrity of religious understandings of the world while embracing the factual basis of evolution, including human evolution, at the same time.

It is equally evident, though, that scientists often misunderstand the public's apprehensions about evolution. That misunderstanding is sometimes magnified by the assumption that religion and science necessarily conflict. For many people, the defining character of human beings is found within religious belief and moral principle; matters of the human soul, afterlife, and suffering, among other vital perspectives about who we are, cannot be considered matters of science open to empirical investigation. Yet the claim that science makes such aspects of religious and spiritual belief unnecessary or wrong, as claimed by some writers, reflects personal philosophy that is indefensible as science. Those who see in the science of evolution a threat to their religious beliefs can thus portray it, and much of science in general, as an offensive philosophy or godless religion rather than as a fulfilling way of finding out new things about the world. There is no doubt that evolution is a material process, which is why its scientific basis is as powerful and believable as the concepts of gravity, continental drift, and the microbial causes of disease. There is no doubt that findings about evolution pose a contradiction to biblical Scripture concerning the literal creation of Earth and all life within a week. Yet these facts need not close communication or thwart efforts to understand how many others have reconciled and embraced the multiple dimensions that enrich people's lives.

In exploring the science of human origins, this book emphasizes that evolution is not the answer to what it means to be human. The answers to this question are as diverse as people. This book instead tries to show how the ongoing investigation of human evolution sheds light on our humanness. Each one of us bears the evidence of the emergence of human characteristics. These characteristics did not originate all at once. There is abundant evidence for the accumulation over time of adaptations related to how we walk, how our brains function, how we interact with our surroundings, and how our social behavior became elaborate. Dramatic change in the environments of human ancestors presented survival challenges, and the benefits of particular adaptations came at a cost. So, for example, the benefits of walking upright ultimately resulted in the disadvantages of back pain, and the enlargement of the brain meant the hardship of giving birth to large-headed babies. A careful look at the great apes shows amazing continuities between these evolutionary cousins and the features that emerged over the course of human evolution. These continuities speak to our primate heritage. Finally, the increasing pace of fossil discoveries has disclosed that our family tree is branching and diverse, like the family trees of virtually all other organisms.

All of these core concepts of evolution are exemplified by the study of human origins. To the investigators and students motivated by curiosity about this topic, these aspects of the scientific story are immensely meaningful and offer many further questions to consider and pursue. With these thoughts in mind, this book is an invitation to explore how certain characteristics of what it means to be human evolved.

This book is an invitation to explore how certain characteristics of what it means to be human evolved.

Milestones in Human Evolution

What makes us unique—different from all other apes, primates, and mammals?

A multitude of physical traits and behaviors define our species. These characteristics did not evolve all at once but took about six million years to accumulate as human ancestors faced many different challenges to their survival over time.

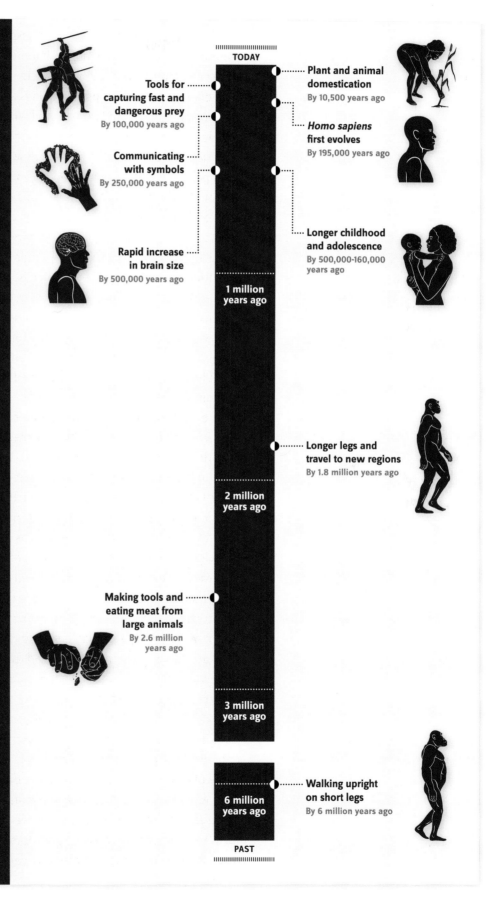

TODAY

Tools for capturing fast and dangerous prey
By 100,000 years ago

Plant and animal domestication
By 10,500 years ago

Homo sapiens **first evolves**
By 195,000 years ago

Communicating with symbols
By 250,000 years ago

Rapid increase in brain size
By 500,000 years ago

Longer childhood and adolescence
By 500,000-160,000 years ago

1 million years ago

Longer legs and travel to new regions
By 1.8 million years ago

2 million years ago

Making tools and eating meat from large animals
By 2.6 million years ago

3 million years ago

6 million years ago

Walking upright on short legs
By 6 million years ago

PAST

PART I

BEING HUMAN, BECOMING HUMAN

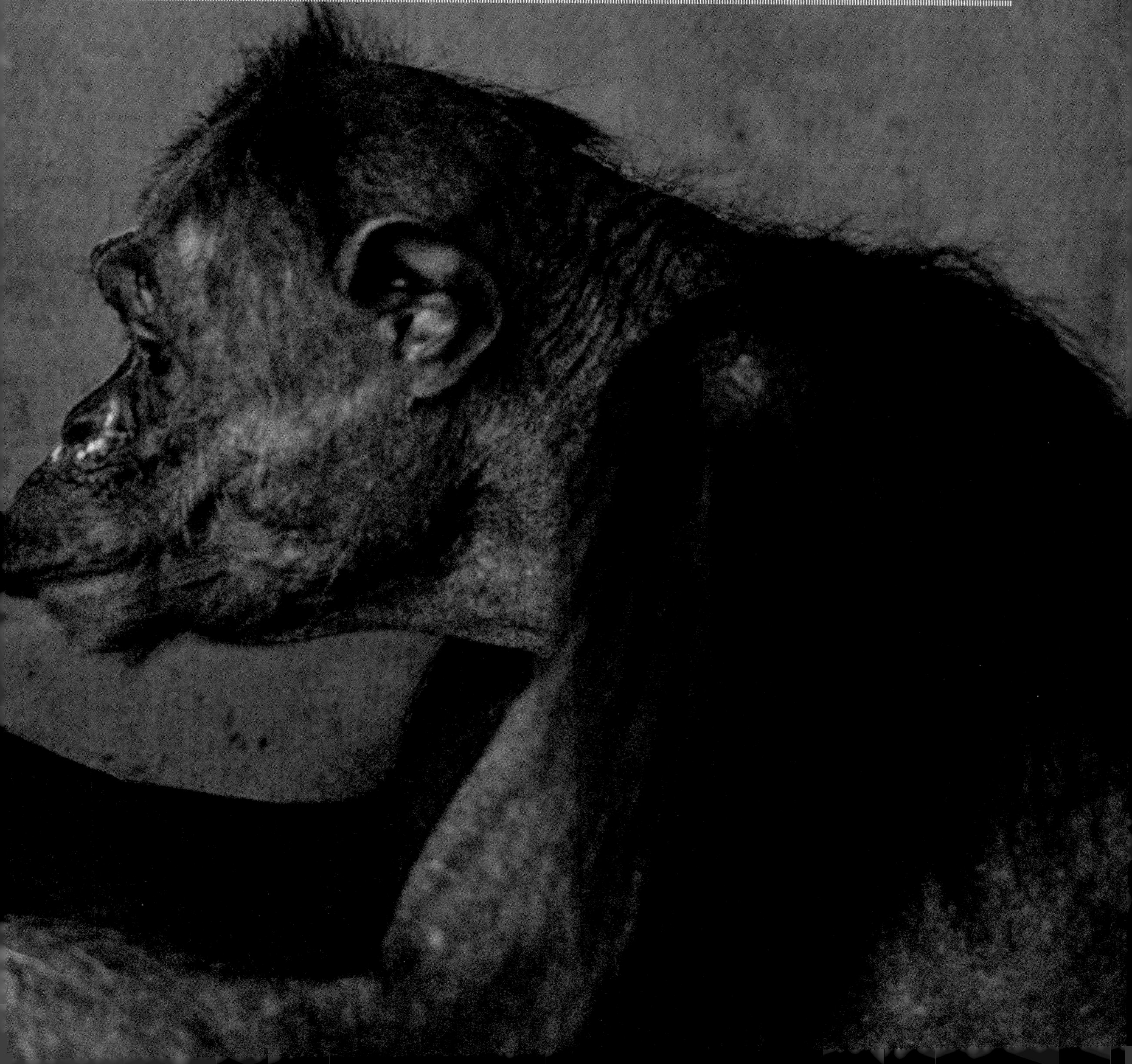

CHAPTER 1

OUR PRIMATE HERITAGE

AS I WATCHED, STILL SCARCELY BELIEVING IT WAS TRUE, I SAW TWO MORE CHIMPANZEE heads peering at me over the grass from the other side of a small forest glade: a female and a youngster," wrote Jane Goodall in her book *In the Shadow of Man.* "They bobbed down as I turned my head toward them, but soon reappeared, one after the other, in the lower branches of a tree about forty yards away. There they sat, almost motionless, watching me."

Previous pages: *Primatologist Jane Goodall's research helped bridge the perceived large gap between apes and humans.*

Opposite: *Bonobos, like this adult female, and their chimpanzee relatives shared an ancestor with humans about six million to eight million years ago. In some aspects of their social behavior, bonobos resemble humans more than chimpanzees do.*

Your ability to pick up this book, read it, learn from it, throw it on the sofa, and tweet about it to a friend all mark you as a primate. Grasping hands, acute 3-D vision, learned behavior, flexible arms, and the urge for social connections are traits we share with lemurs, tarsiers, monkeys, and apes. A suite of additional characteristics helps establish the boundaries that separate primates from organisms that are not primates. Humans are well within primate lines.

The famed taxonomist Carolus Linnaeus named the order Primates in 1758 and placed humans, whom he called *Homo sapiens,* in that group, along with chimpanzees, orangutans, tarsiers, ring-tailed lemurs, and 37 other creatures. It was overwhelmingly evident even then, a century before Darwin published *On the Origin of Species,* that humans are primates. By studying primates, we can not only observe the behavior of our closest kin but also reconstruct how human ancestors might have behaved.

We share numerous features with the 250 species of primates living on Earth. Our hands are able to grasp objects with opposable thumbs, and our fingers are agile because all primates have nails rather than claws. Thus, we can grasp objects, cling to branches or monkey bars, and perform fine manipulations such as grooming or tying our shoes. Compared with other mammals, primates tend to have bigger brains, which they use in social interactions and learning. These behaviors can be as complex as capuchin monkeys showing each other how to use rocks to break open nutritious nuts or a chimpanzee mother teaching her offspring to use sticks to catch ants hidden in the ground. We humans use our big brains to remember thousands of such behaviors. This accumulation of information makes culture possible.

Combining visual signals with a wide repertoire of vocalizations, primates are among the most communicative of animals. Primates' forward-facing eyes bring about creative opportunities for facial expression—grimaces, emotional

Opposite: *Great apes, like this young chimpanzee, take as naturally to the trees as they do to the ground. The earliest members of the human lineage were also good climbers.*

Below: *Fossils and DNA confirm that humans are primates and members of the great ape family. Although we did not evolve from any of the apes living today, we share characteristics with chimpanzees, gorillas, orangutans, and other apes.*

signals, gestures of appeasement and threat. Forward-facing eyes also allow sharp stereoscopic vision and excellent eye-hand coordination, which are useful in capturing prey, judging distances, jumping from branch to branch, and manipulating objects. However, what we gained in vision, we lost in smell. A recent genetic study suggests that the evolution of genes for color vision in primates resulted in the loss of olfactory receptor genes involved in smell. Compared with most other mammals, primates have reduced snouts relative to the size of their braincases.

Primates display an astonishing array of approaches to movement—climbing, leaping, running, swinging, walking, hopping, four legs, two legs, no legs. Humans managed to add swimming and, through technology, flying to the repertoire of

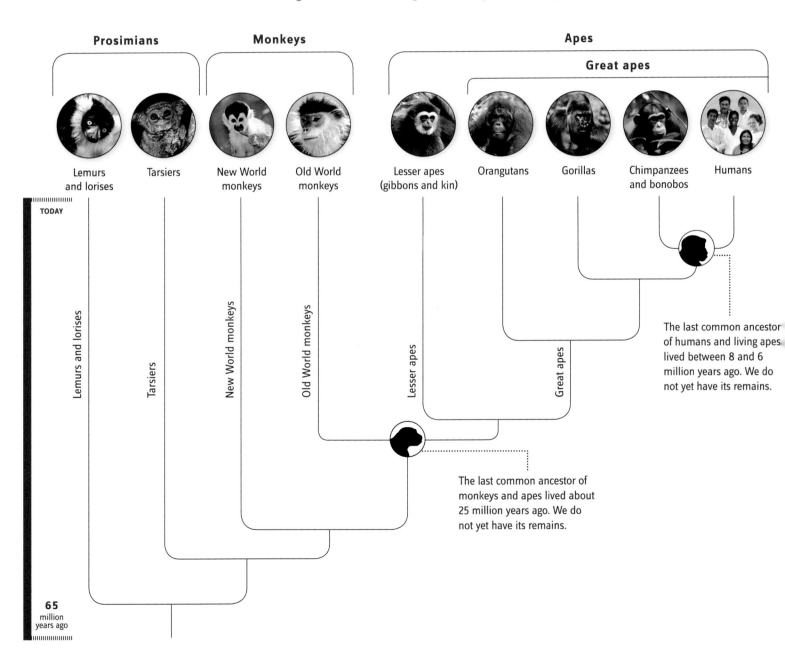

The last common ancestor of humans and living apes lived between 8 and 6 million years ago. We do not yet have its remains.

The last common ancestor of monkeys and apes lived about 25 million years ago. We do not yet have its remains.

primate locomotion. While primates evolved in and still inhabit primarily tropical habitats, their flexible ways of getting around, using their brains and manipulating their surroundings, have enabled them to penetrate and adapt to environments ranging from the hot and humid to the cold and snowy.

Consider the flexibility of our own species; the variety of foods we eat and environments we inhabit; the ways we can walk, climb, run, and jump; our small faces tucked beneath an inflated brain; our expressive eyes and faces; our reliance on vocal communication; our dependence on each other, and on learning, technology, and manipulating the world around us. As wondrous as human qualities are, it's our primate heritage that explains much about our place in the natural world.

A FAMILY OF APES

For more than a century, scientists have noted that the differences between humans and apes are fewer than those separating apes and monkeys. As far back as the 1860s, the anatomist Thomas Henry Huxley observed that humans and apes exhibit marked similarities in overall anatomy as well as in the detailed structure of their teeth, skulls, and other parts. Huxley insisted that, in most details, the brains of chimpanzees and humans differ hardly at all compared with the difference between the chimpanzee's brain and those of monkeys and lemurs; modern neuroscience now confirms Huxley's evaluation. The strong similarities between humans and African great apes, in particular, led Darwin in 1871 to say that Africa was the likely place where the human lineage branched off from other animals'—that is, the place where the common ancestor of chimpanzees, humans, and gorillas once lived.

In recent years, DNA has shown how closely related we are to one another. DNA is the molecular code that makes up an organism's genome; it shapes how an organism grows up and the physiology of its blood, bone, and brains. The level of difference in DNA from one species to the next is a test of how closely or distantly related they are.

While the genetic difference between individual humans is minuscule—about 0.1 percent, on average—studies of the same parts of the chimpanzee genome indicate a difference from humans of about 1.2 percent. The bonobo (Pan paniscus), which is the close cousin of the chimpanzee (Pan troglodytes), differs from humans to the same degree. The difference between human DNA and the DNA of gorillas, another of the African apes, is about 1.6 percent. Most important, chimpanzees, bonobos, and humans all show the same degree of difference from gorillas. A difference of 3.1 percent distinguishes the African apes and us from the Asian great ape, the orangutan. How do the monkeys stack up? All of the great apes and humans differ from rhesus monkeys, for example, by about 7 percent in their DNA.

A thorough comparison of the human and chimpanzee genomes points, in fact, to a wide variety of differences. Segments of DNA that are deleted,

From where?

Did the common ancestor of humans and African apes originate in Africa or Eurasia? Between 23 million and 11 million years ago, apes flourished in both regions. Some scientists consider the Eurasian ape genera Dryopithecus and Ouranopithecus the most plausible candidates for the common ancestor, suggesting that one of them may have followed tropical habitat back into Africa around nine million to seven million years ago as the climate in Eurasia cooled. There the Eurasian ape could have diversified into the lineages that ultimately led to living African apes and humans.

However, other scientists think the African fossil record is still too poor to confidently rule out the possibility that the common ancestor we share with African apes originated and diversified within Africa. Two recent discoveries, Nakalapithecus in Kenya and Chororapithecus in Ethiopia, hint at this possibility. Additional fossil discoveries will continue to shed light on this central question in human evolution.

duplicated, or shifted around in the human genome are not found in chimpanzees, and vice versa. While the similarities can thus be calculated in a variety of ways, they all lead to the same point: Humans, chimpanzees, and bonobos are more closely related to one another than any is to gorillas or any other primate.

From the perspective of this powerful test of biological kinship, humans not only are related to the great apes; we *are* one. The DNA evidence leaves us with

Orphaned infants, a chimpanzee (left) and a gorilla demonstrate a common primate trait: the need to touch, groom, hug, and kiss.

one of the greatest surprises in biology: A wall we've long believed to neatly divide human from ape is not there. The human evolutionary tree is embedded *within* the great apes.

DNA evidence also confirms some of Darwin's most daring ideas about human evolution. In fact, hardly has a scientific prediction so bold, so "out there" for its time, been as thoroughly upheld as the one he made about human evolution beginning in Africa. Ongoing fossil discoveries confirm that the first four million years or so of human evolution took place exclusively on the African continent. It is there that the search continues for fossils at or near the branching point of the chimpanzee and human lineages.

FROM WHEN, FROM WHOM?

Because the degree of genetic difference between two living species can be calculated, so can the timing of their divergence—that is, when their lineages branched away from one another. Primate fossils offer a number of ways to set this "molecular genetic clock." According to the fossil record known so far, 16 million years ago is considered a good estimate of when the orangutan ancestors diverged from the lineage that led to African apes. Since orangutans show about

Monkey (rhesus macaque)

Proconsul

Ape (orangutan)

*The early fossil ape **Proconsul** combined monkeylike and apelike features in its backbone, shoulders, forelimbs, hips, hands, and feet. Like modern apes, it lacked a tail.*

twice the genetic difference from African apes that gorillas show from chimps and humans, it follows that orangutans diverged twice as long ago as the gorilla. The resulting calculation means that the gorilla lineage diverged about eight million years ago from the common ancestor of chimpanzees and humans. This places the divergence between chimpanzees/bonobos and humans at about 6.2 million years ago, which closely matches the fossil evidence for the oldest known human ancestor.

This basic understanding of evolution as a series of branching points from common ancestors helps underline a common misunderstanding. The error is to say that humans evolved from chimpanzees or from monkeys—statements usually set up only to permit derogatory comments about the science and theory of evolution. Although humans are related to all primates and to all living organisms on Earth, we did not evolve from any living primate. We did not evolve from chimpanzees, but rather from a common ancestor that humans shared with chimpanzees around six million years ago. This close relationship puts humans squarely within a group of primates called Hominoidea, or the hominoids. We humans do have our own family tree.

Yet our genealogy is ensconced in an even broader evolutionary tree that includes the living chimpanzees, bonobos, gorillas, and orangutans—and, back through time, all the primates.

CONTINUITY AND UNIQUENESS

Study of apes in their natural habitats and through human interaction has closed many of the gaps we once envisioned separating humans from apes, such as the ability to use tools and symbols. Indeed, dogs can "speak" in a sense, bees perform a kind of wiggle "language," and certain birds use their beaks to manipulate sticks as tools. Yet it's our relationship with the great apes that appropriately frames the unique features of human evolution by enabling us to see those aspects of our humanness that are shared with living apes and inherited from earlier primate ancestors.

Although humans are related to all primates and to all living organisms on Earth, we did not evolve from any living primate.

The noted biologist and human behavior researcher David A. Hamburg put it this way in 1971, commenting on Jane Goodall's pioneering fieldwork with chimpanzees:

The picture of chimpanzee life that emerges is fascinating. Here is a highly intelligent, intensely social creature capable of close and enduring attachments, yet nothing that looks quite like human love, capable of rich communication through gestures, postures, facial expressions, and sounds, yet nothing quite like human language. This is a creature who not only uses tools effectively but also makes tools with considerable foresight; a creature who does a little sharing of food, though much less than man; a creature gifted in the arts of bluff and intimidation, highly excitable and aggressive, capable of using weapons, yet engaging in no activity comparable to human warfare; a creature who frequently hunts and kills small animals of other species in an organized, cooperative way, and seems to have some zest for the process of hunting, killing, and eating the prey; a creature whose repertoire of acts in aggression, deference, reassurance, and greeting bear uncanny similarities to human acts in similar situations.

What Hamburg saw through the lens of Jane Goodall's work holds up for all who spend time with chimpanzees in the field. There are certain physical traits found only in hominoids—the ability to rotate the arm above the head, the lack of a tail, and many others. And humans possess those traits as part of our evolutionary inheritance. As Hamburg observed, however, the continuum between the great apes and our species is found across a wide spectrum of behaviors, emotions, and social interactions. The continuity is what an evolutionary origin predicts. The apishness of humans, and the humanness of apes, continue to be supported by all the evidence that comes with ongoing observations of the primates.

Opposite: Many great ape species lived in Eurasia and Africa (shaded on maps) millions of years ago. As the climate cooled, they began to die out in Europe and Asia.

Ouranopithecus is one of several Eurasian great apes from which African great apes possibly descended.

RISE AND FALL OF THE GREAT APES

The great apes are threatened by extinction. Orangutans are endangered in the only two places where they have survived—the islands of Borneo and Sumatra. The number of western lowland gorillas in Africa has declined by 50 to 70 percent since the early 1990s in areas where they were known to live. Chimpanzees are faring slightly better. Although their geographic range is fragmented and populations are isolated from one another, about 300,000 individuals inhabit more than 50 national parks in 19 African countries. Our other closest living relative, the bonobo, is much closer to extinction.

One has to go back more than 9.5 million years to find an era favorable to apes. Habitat shrinkage and extinction have been the trend since that time—except for the flourishing of the human family tree over the past six million years.

One of the oldest well-documented apes is *Proconsul,* which was represented by several species in East Africa from about 20 million to 18 million years old. *Proconsul* fossils show an intriguing mixture of new traits that foreshadow the later apes, along with traits inherited from older ancestors. Two nicely preserved skeletons found in Kenya show features that typify all later apes, such as mobile hip and ankle joints and the lack of a tail. These are combined with monkeylike features carried on from an ancient common ancestor, such as a long, flexible spine; a narrow chest; and a horizontal, four-legged posture typical of ancient and living monkeys.

Apes that evolved after *Proconsul,* between 16 million and 12 million years ago, show the accumulation of the features that characterize modern apes. Humans inherited from them the impressive mobility of our shoulders, wrists, ankles, and other joints; a broad chest; and a stable lower back. As the ancestors of living great apes evolved more mobile joints, they developed the ability to straighten the elbow as the shoulder rotates. These features enabled great apes to hang and climb along vines and branches, allowing easy movement through the trees below the branches and sometimes on the ground. Similar flexibility later freed human arms and hands to do everything from throwing spears to making tools. Complex manipulation of branches and vines also led these early apes to a knack for bending foliage together to form nests—that is, bedding to

FAQ:
Did humans evolve from chimpanzees?

No. Humans did not evolve from chimpanzees, nor from any other primate living today. But humans and chimpanzees are both primates, and are more closely related to each other than either is to any other living primate. Humans and chimpanzees share a common ancestor, a long-extinct great ape. Fossil and genetic evidence indicates that this common ancestor lived between about eight million and six million years ago. Many similarities that chimpanzees and humans share are part of our inheritance from this earlier ancestor. All apes and monkeys share a more distant relative, which lived about 25 million years ago. It is a misconception, though, that this common ancestor was a monkey, or that humans evolved from monkeys.

sleep on—every night, an intriguing difference between all great apes and other primates.

The heyday of the great apes occurred between 11 million and 9.5 million years ago, a period of astonishing species diversity. An evolutionary explosion produced dozens of species, mainly in the warm, humid, forested environments of Europe and Asia. The shape and structure of their teeth indicate that most apes of this time concentrated on a diet of soft, ripe fruits, although some could handle other foods, including leaves, seeds, and nuts. Soon thereafter, the climate of Eurasia began to cool and show strong differences between wet and dry seasons. Eventually, apes no longer existed in Europe, but only in tropical Africa and the warm forests of East Asia.

With their bigger brains and varied diet, great apes had the ability to survive as forests retreated and the tropical habitats of Africa and Southeast Asia contracted over the past several million years. Human ancestors were the first great apes to sever their dependence on tropical habitats.

HUMAN ORIGINS

While humans are great apes, the great apes are not human. Other apes stand up and walk on two legs from time to time, but we are the only ones who *depend* on walking upright. Other apes take a long time to reach maturity, but we take several years longer and have added many years to our lives by living far past our reproductive years. While great apes use tools, humans are the only ones to use tools to create new tools. Other apes communicate vocally and have a sophisticated capacity to use signals, but only humans *depend* on symbols and language. Other apes are highly social, but only humans communicate with others who live in distant communities.

For students of human evolution, the big questions concern when, exactly, the characteristics unique to humans emerged and what evolutionary path led eventually to the appearance of modern humans, *Homo sapiens*. These questions may seem simple at first. Yet because humans did not appear all of a sudden in fully modern form, but instead went through millions of years accumulating adaptations such as bipedal walking, big brains, and toolmaking, the answers are informed by many lines of evidence and hundreds of discoveries.

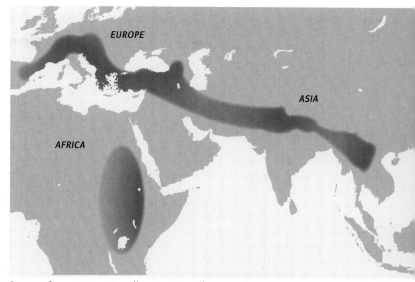

Range of great apes 12 million to 9.5 million years ago

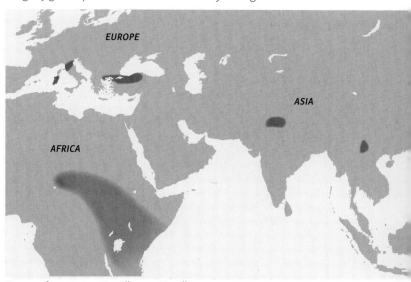

Range of great apes 9 million to 7 million years ago

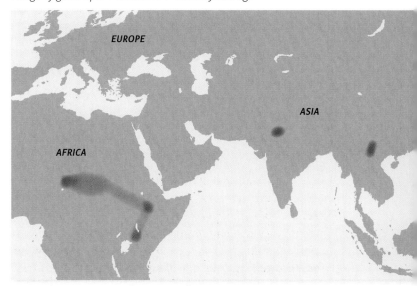

Range of great apes 7 million to 5 million years ago

THE HUMAN FAMILY TREE

SOMEWHERE IN THE ATTIC, MANY OF US HAVE BOXES OF OLD FAMILY PHOTOS. SOME might even hold records of many generations of family ancestors, going back hundreds of years. In old portraits we sometimes recognize eyes, or a nose, or a smile that has been passed down through the generations. Now, think about a half million generations, taking us back seven or eight million years. That's how far back we would have to go to find the last common ancestor our species shared with chimpanzees. The fossil record of humans shows that visible hints of distant relatives are still apparent even across such immense time spans.

The only illustration in Charles Darwin's *On the Origin of Species,* published in 1859, was a tree of life. Darwin felt strongly that a tree was an appropriate metaphor for evolutionary relationships. From common roots, many branches, each representing a different genus or species, could evolve. Darwin also explained how, in the course of evolution, some branches of the tree die off. He did not stress that the tree metaphor might well apply to the human story, although it was implicit. He had enough trouble on his hands suggesting that other creatures had evolved.

We know now that many forms of prehistoric humans lived before us, and that Darwin's tree is as good a model for human evolutionary history as it is for any other life-form on Earth.

Only three years before the publication of *On the Origin of Species,* German limestone workers digging in a quarry in the Neander Valley, not far from Düsseldorf, uncovered a human fossil. It had huge brows, a low forehead, and thick, bowed leg bones. At first, some scientists dismissed the remains as the skeleton of a disease-racked modern human or even a bowlegged Cossack soldier left behind by Russians fighting Napoleon's army. Ultimately, other fossil human bones were found at the site, along with the remains of extinct mammals that indicated the antiquity of these humans, who came to be known as Neanderthals. *Homo neanderthalensis* eventually became the first scientifically described prehistoric human species.

Since the discovery of Neanderthals in Europe, numerous other fossil ancestors have been found. Some finds were met with similar resistance. In 1895, Dutch anatomist Eugène Dubois brought skeletal material of *Pithecanthropus*

Opposite: *Kamada Nakazato, 102, holds her great-great-granddaughter, 4-month-old Yukuzi. Family resemblances often persist through many generations.*

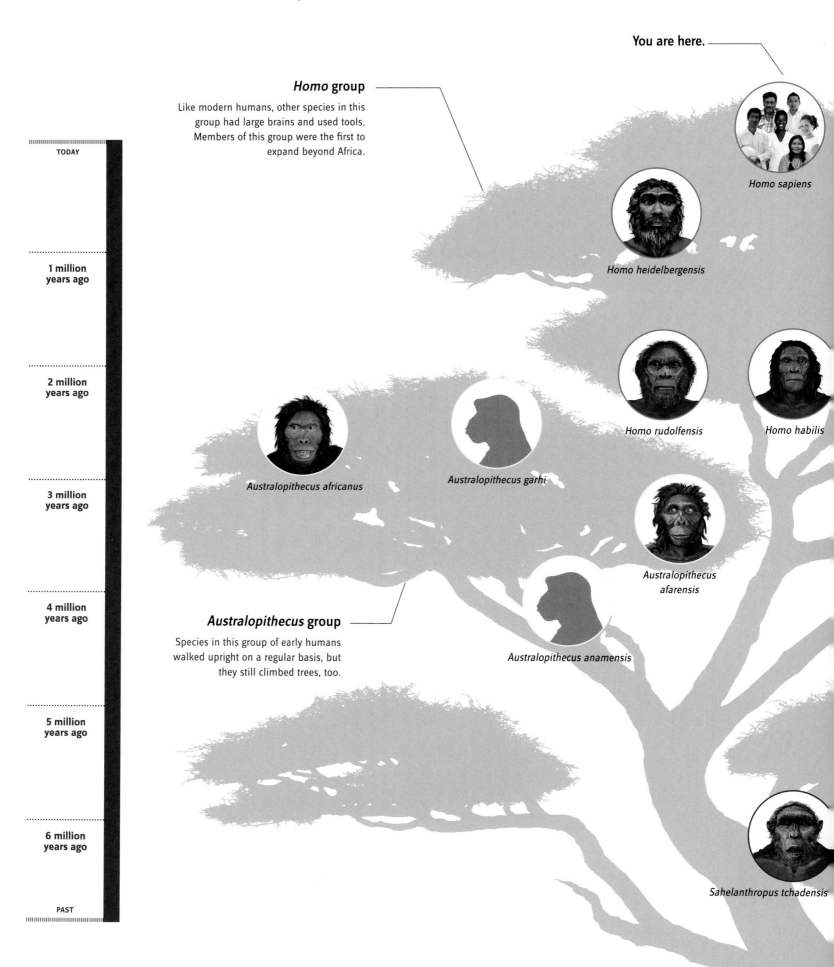

You are here.

Homo group

Like modern humans, other species in this group had large brains and used tools. Members of this group were the first to expand beyond Africa.

TODAY

1 million years ago

2 million years ago

3 million years ago

4 million years ago

5 million years ago

6 million years ago

PAST

Australopithecus group

Species in this group of early humans walked upright on a regular basis, but they still climbed trees, too.

Homo sapiens

Homo heidelbergensis

Homo rudolfensis

Homo habilis

Australopithecus africanus

Australopithecus garhi

Australopithecus afarensis

Australopithecus anamensis

Sahelanthropus tchadensis

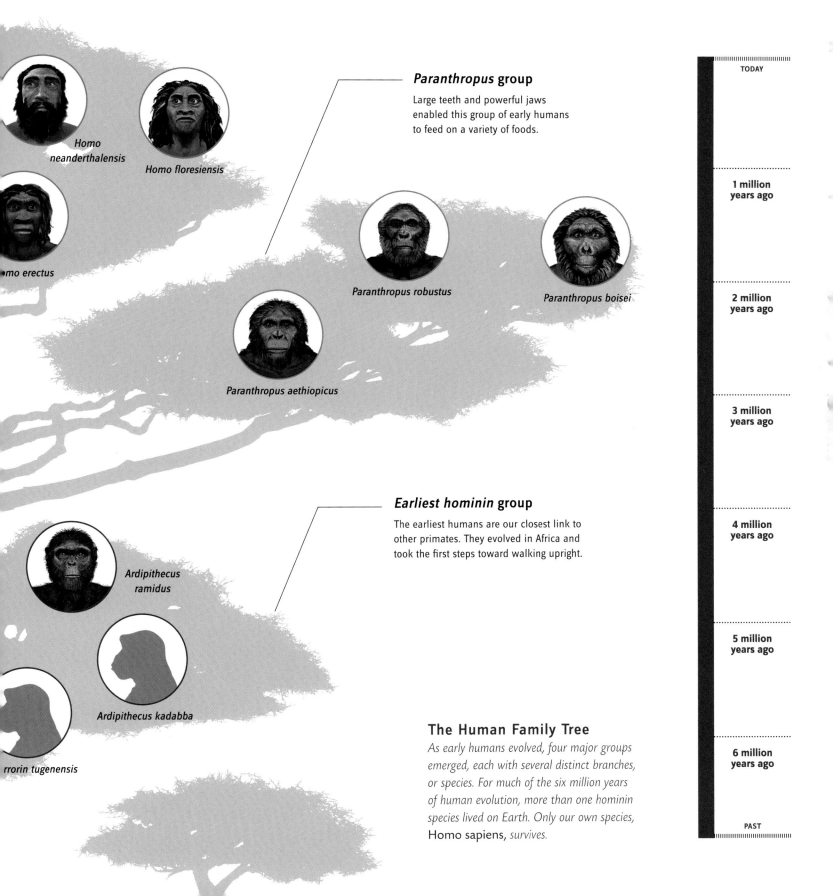

Paranthropus group

Large teeth and powerful jaws enabled this group of early humans to feed on a variety of foods.

Homo neanderthalensis

Homo floresiensis

mo erectus

Paranthropus robustus

Paranthropus boisei

Paranthropus aethiopicus

Earliest hominin group

The earliest humans are our closest link to other primates. They evolved in Africa and took the first steps toward walking upright.

Ardipithecus ramidus

Ardipithecus kadabba

rrorin tugenensis

The Human Family Tree

As early humans evolved, four major groups emerged, each with several distinct branches, or species. For much of the six million years of human evolution, more than one hominin species lived on Earth. Only our own species, Homo sapiens, *survives.*

TODAY

1 million years ago

2 million years ago

3 million years ago

4 million years ago

5 million years ago

6 million years ago

PAST

What's in a name?

Until the late 20th century, members of the human lineage enjoyed an elevated taxonomic status in a family, Hominidae, dedicated exclusively to them. This status was undermined by genetic studies suggesting chimpanzees and humans are more closely related to each other than chimpanzees are to any of the other great apes. This was a shocker that led taxonomists to reconsider the classification of apes.

Many researchers accept the reshuffling and renaming of species to reflect the new genetic evidence. Humans are now placed in a grouping known as a tribe (not a family) named Hominini, informally known as hominins. This name applies to our entire evolutionary group, including fossil species, since the time of our divergence from the common ancestor we share with chimpanzees. Our closest living biological relatives—chimpanzees and bonobos—are in a sister tribe, Panini. And the family Hominidae, or hominids, now includes not only the human lineage but also chimpanzees and all the other living great apes.

erectus (now *Homo erectus*) from Java, only to encounter enormous skepticism over claims that he had found the so-called missing link. In the 1920s Australian anatomist Raymond Dart's claim that the Taung Child fossil from South Africa was a human ancestor also met with disbelief. Critics claimed it was a juvenile gorilla. It would take another 20 years, during which new fossil discoveries were made, for the scientific world to accept the 2.8-million-year-old Taung Child as an ancestral human species, *Australopithecus africanus*.

The report in 2004 of a newly discovered species of human that lived as recently as 17,000 years ago on the Indonesian island of Flores again stirred controversy.

Some scientists were initially reluctant to accept *Homo floresiensis* as a species in our family tree not only because it was so recent but also because it was so small, standing just over 1 meter (39 inches) tall. It also had a brain so tiny that nothing like it had been seen in the human lineage for more than two million years. Supporters of naming it a new species thought an ancient relative, perhaps *Homo erectus,* had become marooned on an island and evolved to a smaller size, as often happens with island species. Detractors called it a modern human that suffered from genetic abnormalities, such as microcephaly and dwarfism.

The variation we see in prehistoric human fossils can be both confusing and provocative. New discoveries often shock us, or at least raise questions. How closely related are we to these species? Did we coexist with them? Were there encounters between species? The answers to many of these questions can be found by looking at the genealogy and chronology of the human family tree.

It is not the existence of an evolutionary family tree that is in question today, but rather its size and shape. The number of branches, representing genera and species, and their position are much debated among researchers and are further confounded by the admittedly incomplete fossil record. This debate is sometimes perceived as uncertainty about evolution, but that is far from the case: It concerns the precise evolutionary relationships—essentially, "who is related to whom."

The fossil evidence now shows that there were periods when three or four species of early humans lived at the same time. There is also consensus among researchers that there was tremendous variation in the skulls, teeth, and bodies in the human lineage over millions of years. The big debates mainly concern exactly how to divide those variations into distinct species. Nonetheless, we know now that the human family tree has many more branches and deeper roots than we knew even a couple of decades ago.

One way to begin to understand our complex genealogy is by looking at the family tree as having four major parts: one for the earliest members of our lineage, and one each for members of the genera *Australopithecus, Paranthropus,* and *Homo.*

THE EARLIEST HOMININS

The earliest hominins—members of the human family tree that lived between seven million and four million years ago—composed a group we still have much to learn about. Fossils of human ancestors of this age are extremely rare; none were known until the 1990s. Since then three genera have been named. Mystery surrounds these hominins not only because of the rarity of specimens but also because researchers interpret these fossils differently.

Scientists have not yet discovered the earliest hominin species. The oldest members of our lineage were surely quite apelike, but they would most likely have shown some adaptation to a lifestyle different from that of other great apes. The two oldest traits that distinguish the human lineage from other apes' are upright posture, related to walking on two legs, and reduced canine teeth in males, which could relate to changes in social life (as we will see in Chapter 5). The presence of these features defines members of the Hominini, our evolutionary group.

We do not yet know which of these adaptations was the first to evolve in our lineage. As fossils are found that bring us ever nearer to the actual time when the human and chimpanzee lineages diverged, the more difficult it will likely be to figure out which lineage is which. Currently, the earliest known hominin is *Sahelanthropus tchadensis,* meaning "Sahel man from Chad." It is known from a chimpanzee-size skull, two mandibles, and assorted teeth. These fossils were first found in the Djurab Desert of northern Chad in 2001, far from areas in southern and eastern Africa that have earned distinction as "cradles of humanity." *Sahelanthropus* lived between seven million and six million years ago, during a

Below: *Layers of desiccated earth, hundreds of meters thick, come to light in the Afar Depression of Ethiopia. Erosion and excavations here reveal one of the most complete records anywhere of human evolution.*

Lifelike reconstructions of early humans, created by artist
John Gurche, combine the latest forensic techniques, fossil
discoveries, and 20 years of experience.

Australopithecus africanus

A. afarensis

Paranthropus boisei

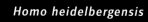

Homo heidelbergensis

H. neanderthalensis

H. erectus

H. floresiensis

Australopithecus africanus, STS 5
About 2.5 million years old
Small braincase; sloping face

Homo rudolfensis, KNM-ER 1470
About 1.9 million years old
Braincase larger than in earlier humans; sloping face

Homo erectus, Sangiran 17
About 1 million years old
Medium braincase; distinct brow ridge

period known as the Miocene and very soon after the split between chimpanzees and the human lineage. Its presence in Chad also shows that early hominins were more widespread in Africa than previous evidence suggested.

The key features that signal the status of *Sahelanthropus* as a member of the human family tree include canine teeth that were small compared with those of earlier Miocene apes, along with the position of the foramen magnum, the opening in the base of the skull through which the spinal cord connects to the brain. In *Sahelanthropus,* as in other hominins, it is in a more forward position than it is in great apes. This placement suggests that *Sahelanthropus*'s neck was oriented vertically—a strong clue that *Sahelanthropus* held its body in an upright position and may have walked upright.

The two other genera of this early group of hominins, *Orrorin* and *Ardipithecus,* also lived in the late Miocene. Like *Sahelanthropus,* they possessed uniquely human traits mixed with apelike ones.

The finds of *Orrorin tugenensis,* from central Kenya, are between 6.2 million and 5.8 million years old. The species is known from an assortment of bone fragments representing several individuals. These fragments include a femur that shows a muscle groove associated with bipedal locomotion, along with thickening of the bone bridge, known as the femur neck, which connects the femur to the hip socket—again, strongly suggestive that this species was adapted for walking on two legs.

Two species of the genus *Ardipithecus* are known; specimens representing dozens of individuals have been found in the Afar Depression of Ethiopia. The earliest known *Ardipithecus* species, *Ardipithecus kadabba,* lived between 5.8 million and 5.2 million years ago. *Ar. kadabba* had the most apelike canines yet seen in the human lineage, according to the researchers who studied it. *Ar. ramidus,* which lived about a million years later, had smaller canines.

A partial skeleton of a female *Ar. ramidus,* known as Ardi, which was described in 2009, shows that this hominin moved both by climbing and by walking upright. Although Ardi had long arms and hands, she didn't use them for hanging from branches or walking on her knuckles as living great apes do. Instead, she moved in the trees using the palms of her hands and a grasping big toe. Ardi's upper pelvis, which was shorter and broader than an ape's, indicates that she also walked bipedally. Her teeth had thicker enamel than do the teeth of apes that eat ripe fruit, yet her molars were not as expanded as those of later hominins that ate a diet of abrasive foods. Ardi's species probably used its arboreal and terrestrial lifestyle to exploit a wide range of foods.

There is little question that the fossils assigned to these three genera are relevant to the story of the evolutionary transition from apes to humans. Yet the question of how these three lineages are related to the last common ancestor, to each other, and to later members of our lineage remains unanswered. At one time or another, it has been argued that each of these genera actually represents an ape that does not belong to our evolutionary group, the hominins. It has also been suggested that the three genera are actually one and the same hominin. This makes for mystery and excitement at the base of the human family tree as researchers try to discover which one of these genera may have evolved into the australopiths.

THE AUSTRALOPITHS

The australopiths, a highly successful group of hominin species, spread from southern and eastern Africa to north-central Africa and thrived between 4.2 million and 2 million years ago. Australopiths were chimpanzee size, like earlier hominins, and had chimpanzee-size brains. Their feet, legs, pelvises, spines, and heads all show that they were committed to walking upright. Their legs, however, were short relative to the rest of the body, while their arms were long and well muscled—which suggests that in addition to walking bipedally on the ground, they were comfortable climbing trees.

The teeth of australopiths show that the trend toward smaller canine teeth seen among earlier hominins continued in this group. In contrast to earlier hominins, however, the australopiths had large cheek teeth, a condition known as megadontia, and thicker enamel. These were most likely adaptations to tougher, more abrasive foods, such as fibrous plants, than the softer fruits, such as figs, that may have been more common in their ancestors' diet.

The earliest known member of this group, *Australopithecus anamensis,* lived from at least 4.2 million to 3.9 million years ago, based on Kenyan and Ethiopian fossils. It is a likely descendant of the genus *Ardipithecus. Australopithecus anamensis,* in turn, is a probable ancestor of *Australopithecus afarensis,* the group to which the hominin superstar "Lucy" belongs, along with the world's oldest fossil skeleton of a young early human individual, a three-year-old from Dikika, Ethiopia, who lived about 3.3 million years ago.

The australopiths survived for more than two million years, about one-third of the entire human past, and most likely produced the genus *Homo.* This means that there is a bit of australopith in each of us; it is easy to see in our upright posture, our S-shaped spine, our forward-facing toes, and our broad knees.

THE "NUTCRACKERS"

The third major branch of the human family tree was successful for one and a half million years but arrived at an evolutionary dead end. This was the genus *Paranthropus,* which includes three species that were close cousins of ours. The earliest appearance of this genus was around 2.7 million years ago. Highly specialized teeth and jaws evolved in this group, allowing individuals to exert tremendous chewing forces on any foods placed between their very large cheek teeth. The thick enamel that capped these teeth, along with enormous chewing muscles that connected from the jaw to a crest on top of the skull, probably served them well when eating tough foods, such as tubers, which take a long time to chew.

When paleoanthropologist Mary Leakey discovered *Paranthropus boisei* in 1959, her husband, Louis, referred to it as "nutcracker man" because its robust

Opposite and below: *Fossil skulls show that modern* **Homo sapiens** *evolved from earlier humans. A skull from Fish Hoek, in South Africa, represents modern humans.*

Homo heidelbergensis, *Petralona*
About 350,000 years old
Large braincase; large brow ridge

Homo sapiens, *Fish Hoek 1*
About 4,800 years old
Largest braincase; flat face directly under forehead

teeth and jaws seemed adapted for heavy chewing and grinding. New research shows that the southern African species *P. robustus* also ate insects and possibly small mammals. Their survival strategy, however, might have been to rely on eating tough and hard foods, such as tubers and seeds, when food was scarce. This strategy put them in competition with many other large mammals—especially ancient species of baboons and pigs, which also favored some of these same foods. *Paranthropus boisei* flourished in East Africa after two million years ago.

Excavations at the Drimolen site in South Africa produced the remains of **Paranthropus robustus**. *The quest for knowledge of human ancestry often includes rigorous archeological excavations and painstaking analysis of finds.*

Yet by around 1.2 million years ago, all *Paranthropus* species had become extinct.

Considerable debate surrounds the *Paranthropus* and *Australopithecus* branches of our family tree. Some researchers see the chewing adaptation of *Paranthropus* as a trait that clearly distinguishes it from australopiths. Others, however, suggest that *Paranthropus* species are not different enough from australopiths to warrant designating them a separate genus. This is why the *Paranthropus* hominins are sometimes referred to as "robust australopiths" and are grouped on the tree using the *Australopithecus* genus name. One might read about *Australopithecus boisei*, for example, but it is the same creature as *Paranthropus boisei*. Other researchers do not see the "nutcrackers" as a unified group at all. They see a southern African species, *Australopithecus (Paranthropus) robustus,* that evolved out of a southern australopith, *Australopithecus africanus.* In this scenario, the other big-toothed hominins evolved from an East African australopith, possibly *Australopithecus afarensis.*

THE GENUS *HOMO*

It was not from a big-toothed *Paranthropus* but most likely from a more delicate-featured species of *Australopithecus* that the genus *Homo* evolved. It is to this fourth major group on the human family tree that our own species, *Homo sapiens,* belongs. *Homo* is the most widespread, diverse, and recent of all the hominin genera. It is also the genus known for its enlarged brain and toolmaking.

Homo is distinguished from the australopiths by several key traits. In general, the brains of *Homo* species are larger than those of other hominin genera. The faces of *Homo* are usually smaller and less projecting than earlier hominins'. In our own species, the decrease in facial size reached an extreme: We are the only mammal whose face is situated beneath the front part of the brain, and nearly half of our vertical face is made up of the forehead. Earlier species of *Homo* typically had a distinct bony ridge over their eyes, called a brow ridge. *Homo*'s cheek teeth are significantly smaller than the australopiths', suggesting a change in diet or in how foods were processed. The skeleton of *Homo* is revamped such that the legs are proportionally longer than those of australopiths, and the arms are proportionally shorter. This revamping hints at further specialization as bipeds and a commitment to life lived on the ground. *Homo* is frequently associated with stone tools, and such tools continue in the archeological record well after all hominins other than *Homo* became extinct.

Because of *Homo*'s association with tools, Louis Leakey suggested that the presence of tools alone was enough to show that *Homo* was present. This would place *Homo* as far back as 2.6 million years ago. However, the oldest known molar teeth similar to those of the earliest known *Homo* species, *Homo habilis,* along with an upper jaw from Ethiopia assigned to this species, are currently no older than 2.4 million to 2.3 million years old.

After two million years ago, *Homo* became a wide-ranging genus that included between 6 and 12 species. As far as we know, it was a species of *Homo*—*Homo erectus* or something very similar to it—that was the first hominin to disperse beyond Africa. This raises questions as to whether certain samples of *Homo* fossils found inside and outside Africa deserve to be called separate species. The case of fossils from the site of Dmanisi, in the Republic of Georgia, illustrates the point.

Dmanisi is the source of an amazing trove of early *Homo* fossils, dated between 1.78 million and 1.75 million years ago. The fossils at Dmanisi are consistent in many ways with African *H. erectus,* yet they show some traits, such as a proportionally small brain and short stature, that suggest they might be an earlier form of the genus *Homo.* These early emigrants had adapted to an environment far north of their African homeland, where they encountered new climates, plants,

How old is it?

Scientists have developed many methods to determine the ages of fossils, early tools, and ancient sediments. All of these methods focus on properties that change over time. For example, potassium-argon, argon-argon, carbon-14 (or radiocarbon), and uranium-series methods measure the amount of radioactive decay of chemical elements; the decay occurs in a predictable manner over long periods of time. Thermoluminescence, optically stimulated luminescence, and electron spin resonance methods measure the number of electrons that are absorbed and trapped inside a rock or tooth over time. Paleomagnetism compares the direction of the magnetic particles in the ground with worldwide shifts in the Earth's magnetic field, which are dated and calibrated using other dating methods. Faunal dating estimates the age of sediments based on fossil animals with geologic ages known from other sites. Because these methods are based on different chemical, physical, and biological principles, scientists cross-reference results to confirm their accuracy.

and animals. Some scientists proposed a new name, *Homo georgicus,* to account for the distinctive features of the Dmanisi fossils. The question is whether *H. georgicus* was different enough from *H. erectus* to warrant species status.

The answer to this question is partly a matter of how much variation one accepts within a fossil species. Variation within a species is common and expected. The variation in living populations of chimpanzees and our own species clearly shows this. Without the benefit of observing mating habits among fossil species, however, paleoanthropologists are left with the physical variation visible in bone, the distribution of fossils in time and place, and inferred adaptations to different environments to make the case for or against the designation of new fossil species. It often takes many years and an abundance of fossils from multiple times and places before scientific debates such as the one surrounding the Dmanisi fossils can be resolved.

It took 150 years to resolve debates about the relationship of Neanderthals to modern humans. As recently as the 1980s, modern humans and Neanderthals were considered by many researchers to be subspecies of *Homo sapiens.* They were called *Homo sapiens sapiens* and *Homo sapiens neanderthalensis,* respectively. Today, we can look not only at the bones but also at the genes of these two hominins through new techniques developed to extract ancient DNA from Neanderthal remains. Based on all available lines of biological evidence, we are a separate species from Neanderthals but share a common ancestor in *Homo heidelbergensis,* a species that evolved from *Homo erectus* by at least 700,000 years ago. According to the genetic data, the lineages of *H. neanderthalensis* and *H. sapiens* diverged from this common ancestor between 400,000 and 350,000 years ago. The question of why the Neanderthals became extinct while our species proliferated remains among the most fascinating questions to answer about human evolution.

Modern humans, *Homo sapiens,* arose by 200,000 years ago, and thus our species has been around for only 3 percent of the entire period of human evolution. Compared with our hominin predecessors and the many species that thrived for much longer times than ours, we are newcomers. What happened to our ancestors and their vanished lifeways? Are we vulnerable as well? What could account for our current state as a single species that has spread all over the world, the lone representative of a once diverse evolutionary tree? New information from the environmental sciences about ancient climates may provide some answers.

FAQ:
Is there a missing link?

A "missing link" implies a linear progression from one species to another. But evolution occurs in branching patterns, making a multibranched tree a more appropriate model for understanding evolutionary relationships.

Fossils representing transitional species are not at all rare, especially given the rarity of fossils in general compared with the diversity of life that once existed. In fact, all fossil species represent a transitional form. Like all living species, they possess a combination of unique traits and features inherited from distant ancestors.

Humans and chimpanzees have been evolving independently for millions of years. A missing link, a creature halfway between a human and a chimpanzee, is therefore not a useful concept.

SURVIVAL OF THE ADAPTABLE

"WHEN I FIRST EXCAVATED IN THE RIFT VALLEY, IT WAS COMMON KNOWLEDGE THAT human ancestors emerged on the dry African plains," wrote Rick Potts about his work on Olorgesailie in Kenya. "Yet scrambling up even a single gully, I couldn't help but notice the evidence of vast change over time in the layers of that eroded landscape. Above the white silts of an ancient lake was the brown soil of a dry environment, covered by a gray ash violently spewed from a nearby volcano; then the lake returned, followed by a hard white line when the waters dried up completely. Was it the constant survival challenge of the savanna—or was change itself the more potent force behind the defining qualities of our species?"

Opposite: *Molten lava fills the sky above a volcano in Iceland. Violent natural events, now as over millions of years past, signal our planet's dynamic nature and remind us that human evolution occurred in conjunction with ever changing environments.*

The living world is a display of astonishing adaptations. These adaptations embrace all the structures and behaviors that have favored the survival and reproduction of organisms in the times and places in which they evolved. Powerful claws and a long, sticky tongue do a lot to assist an anteater in digging up and capturing ants. The short "flippers" of penguins are useless for flight, but along with the birds' insulated, bullet-shaped bodies, they help them catch fish in icy Antarctic water. The idea of adaptation extends also to behavior and interactions with other species. The African honeyguide, for example, possesses a keen instinct for finding bee nests; while the honey badger, following the bird, is capable of ripping open the nests to get to the honey, which both honeyguide and badger feed upon.

Over time, a population of organisms evolves in response to the challenges and opportunities of its environment. As grasslands expanded across Africa, prehistoric antelopes evolved teeth that could efficiently chew tough blades of grass that grew on the plains. As the grazing herds expanded, flesh-eating cats became fast, effective killers, and hyenas evolved powerful jaws to crack open the nutritious marrow bones that no other carnivore could break.

Our bipedal cousins were no different. In the lineage of *Paranthropus boisei* the molar and premolar teeth became larger over time, and powerful jaw muscles focused the force on these teeth in a way that favored chewing tough

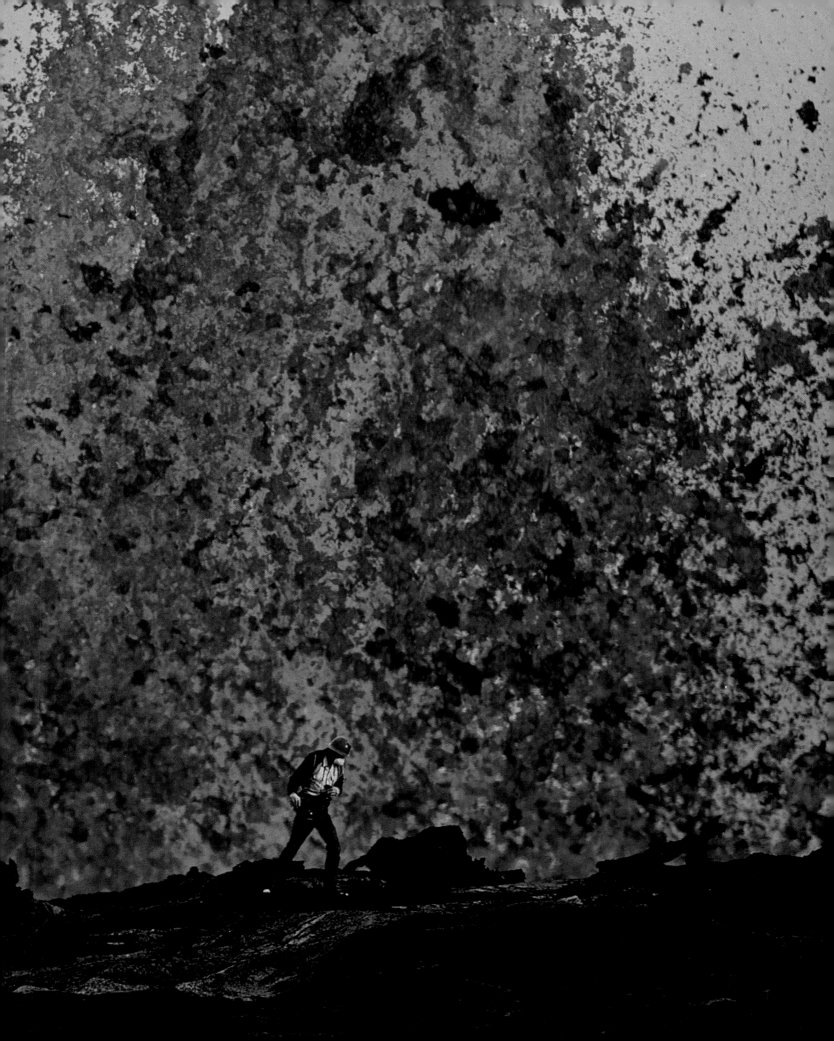

A Nile crocodile attacks a wildebeest crossing a river in Kenya. Our ancestors were vulnerable to many of the same dangers as other creatures were.

and abrasive foods. Considerably later, the short extremities and broad bodies of *Homo neanderthalensis* helped conserve heat and served as adaptations to the cold conditions of Europe in which this species initially evolved.

One of the basic principles of biology, therefore, is that adaptations emerge as organisms face the ongoing tests of survival in their surroundings—finding food, avoiding predators, attracting mates, warding off the cold, and locating shelter.

ADAPTIVE CHALLENGES

Our species, *Homo sapiens,* is recent on the evolutionary scene, having first appeared only about 200,000 years ago. Although all earlier hominins are now extinct, many of their adaptations for survival—an appetite for a varied diet, making tools to gather food, caring for each other, and using fire for heat and cooking—make up the foundation of our modern survival mechanisms and are among the defining characteristics of our species.

Life was not easy for our ancestors. Without claws or canines, the earliest hominins were physically more or less defenseless. Like other primates, they could probably toss rocks, wave sticks, and create a big fuss when threatened. They probably slept in trees at night or, at a minimum, huddled together in groups on the ground.

The hunting and scavenging efforts of later hominins, including the first to make stone tools, brought them close to animals that could injure them. They were meals for crocodiles and hyenas, as well as for big cats. Flash floods, volcanic eruptions, droughts, and other natural disasters added to their tribulations.

Early humans ultimately developed ways to cope with such dangers, but predators and scavengers always lurked. The lower jaws and limb bones of ancestors who lived between 4.4 million and 3 million years ago are often marred by the gnawing of dangerous carnivores. At the later site of Olorgesailie, where handaxes were left behind by the thousands, no early human remains came to light despite decades of searching. Then researchers were struck by the possibility that early humans at this site may have found safety in the highlands at night, when predators typically prowl near water holes. Acting on this hunch, the research team began to dig in the upper margins of the lowlands, and right away found a fossil human cranium that was 900,000 years old. The only pieces that remained were bits of the braincase and the brow ridge, which bore puncture marks from the teeth of a carnivore. This early human never made it home.

Fossil human remains evince clues to other dangers, including illness. One example is a *Homo erectus* adult female whose skeletal remains, found at East Turkana, Kenya, were covered in a layer of abnormal bone. Researchers diagnosed a painful condition in which her bones essentially bled, caused by a disease associated with an overdose of vitamin A. How could this have happened? It turns out that the livers of carnivorous animals concentrate this vitamin at a level extremely toxic to humans, who fall victim to this terrible condition when the liver of a predator is accidentally consumed. Eating meat and even killing carnivore competitors were survival strategies of our ancestors, but in this case, a small mistake proved deadly.

Without medicine, even minor infirmities could be fatal. The robust appearance of the Kabwe cranium from Zambia belies the possibility that this *Homo heidelbergensis* individual died from a small but fatal infection. This individual is one of the oldest known to have had tooth cavities, including ten that invaded the upper teeth. A small perforation in the temporal bone of the skull leads to a larger pit on the interior and shows that either dental disease or a chronic ear infection was the cause of death.

What about fossil evidence of murder or even warfare? Wooden spears about 400,000 years old are preserved, and stone spheroids that could have been thrown date back nearly 2 million years. However, there is no evidence that multiple hominins ever died at the sites where they have been found. The earliest known death from a sharp stone point occurred in one of several Neanderthal individuals buried during thousands of years in the Shanidar Cave of northern Iraq, dated between about 45,000 and 35,000 years old. A severe wound to one of this individual's ribs resulted from a forceful thrust of a stone tip from the left side. Before this, there is no sign of intentional injury in the fossil record. Multiple

Bite marks

Bite marks similar in shape to those made by the teeth of living crocodiles remain in the ankle bone of a partial foot, suggesting the fate of a 1.8-million-year-old **Homo habilis**.

deaths in one place as the result of warfare occurred only more recently and are associated exclusively with our own species.

APPROACHES TO SURVIVAL

With all the signs of injury, disease, and death, one might wonder how our early ancestors survived at all. But this gives the wrong picture. Although early hominins may have been relatively defenseless from a physical standpoint, part of their primate heritage included impressive defenses against predators, including being social and vocal. Primates in social groups keep watch over each other. Together, they can stay aware of predators and may gang up to scare them away. Many primates have warning calls as well; some are specific to attacks from birds, snakes, and leopards. Social and vocal behaviors like these may have made it reasonably safe for our ancestors to venture away from the trees in the first place.

Throwing stones and using weapons were important additions to self- and group defense. There is also clear evidence that by about 800,000 years ago some hominins had controlled fire and built well-defined hearths. Most animals have an innate fear of fire and would be likely to have kept their distance from a campfire. The light from the campfire would also have helped our ancestors avoid surprise attacks. Between 800,000 and 400,000 years ago, when we see strong evidence for building both shelters and hearths, the level of human social cooperation had reached a crucial milestone.

Earth's Changing Climate and Human Evolution

Earth's climate has fluctuated between warm and cool over the past ten million years. The ratio of two oxygen isotopes, as measured in cores drilled from the ocean bottom, ranges from about 2.5 to 5.0 parts per million. This measure reflects both worldwide ocean temperature and the amount of glacial ice. Particularly dramatic fluctuations marked the six-million-year period of human evolution.

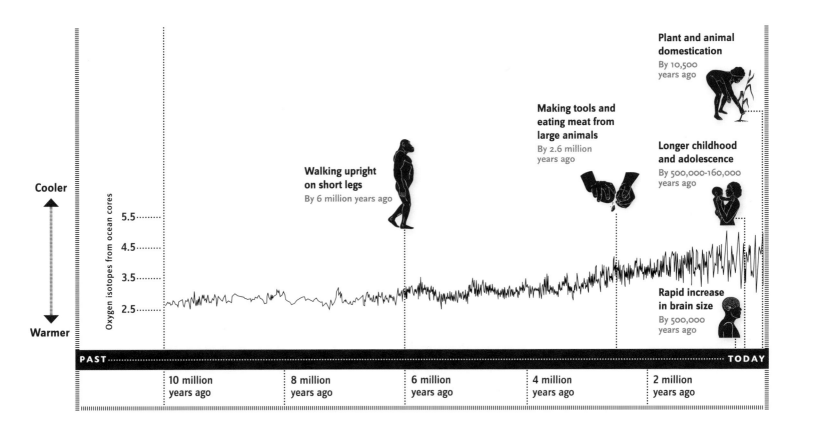

Plant and animal domestication
By 10,500 years ago

Making tools and eating meat from large animals
By 2.6 million years ago

Longer childhood and adolescence
By 500,000-160,000 years ago

Walking upright on short legs
By 6 million years ago

Rapid increase in brain size
By 500,000 years ago

Cooler

Warmer

Oxygen isotopes from ocean cores

5.5
4.5
3.5
2.5

PAST   TODAY

10 million years ago | 8 million years ago | 6 million years ago | 4 million years ago | 2 million years ago

Possibly the ultimate evidence that social support had become an essential part of the human survival package is purposeful burial. Instead of leaving the dead where they lay, Neanderthals and modern humans buried their dead. It could be argued that this was simply a sanitary measure, but there is evidence that suggests far more was going on. We can see this at Shanidar Cave, where an adult male Neanderthal was carefully placed on his side in a shallow grave in a fetal position. There is good evidence that colorful flowers and evergreen boughs were intentionally placed in the burial with him.

Since the physical remains of the deceased do not benefit from burial goods or rituals, did mortuary practices develop mainly as an advantage to the living? The significance of the earliest burials, at least 100,000 years ago, may be twofold. First, rituals reinforce social bonds and may have helped earlier human groups cope with life's difficulties by allowing grieving. Second, burials show that humans were able to conceive of something other than the immediacy and harsh realities of their daily lives. Perhaps they imagined an afterlife for their departed loved ones. Or perhaps a better future for themselves.

ADAPTATIONS AND ADAPTABILITY

The causes of death offer clues to the events of natural selection that helped shape our human ancestors' adaptations. But there was another, equally pervasive influence on evolution—the uncertainties of climate change. While predation and disease exerted an ever present risk, uncertainties in the supply of food and other necessities also posed an incessant challenge to survival.

What happened as early humans tried to adjust to weather variations from one season to another or to wider swings in rainfall and temperature over time? And what about the misfortunes visited upon them by monsoons, intense droughts, or massive volcanic eruptions?

A species depends not only on the particular way of life that best matches its surroundings but also on keeping certain options open and adjusting to whatever trials or opportunities occur as things change. The process of evolution not only shapes adaptations to specific habitats but also shapes ways of life that confer a certain degree of *adaptability.* Since individual species may persist for hundreds of thousands, sometimes millions, of years, the capacity to adjust to novel situations is an important outcome of evolution, offering resilience and an ability to recover from difficult times.

This idea of adaptability has now become a pillar in the understanding of human origins. One reason is that our own species, *Homo sapiens,* may well be the most adaptable mammalian species ever to evolve on Earth. Just look at all the places on the planet where humans live today, at our capacity to alter our surroundings to suit our tastes, and at our propensity to seek out completely novel and challenging places to visit, even outer space.

Scientific research shows now that the period of human evolution has been one of the most volatile eras of environmental change in our planet's history.

A dried-up Texas lake bed may resemble the remnants of African Rift Valley lakes that disappeared during prolonged droughts.

Evidence of vast swings between wet and dry, and between warm and cold, casts our own survival story in a new light.

ERRATIC ENVIRONMENT

Some time ago, any question concerning the environment in which humans evolved seemed entirely resolved: Early humans were adapted to the African savanna. Walking upright on two legs and making implements were critical to the survival of ancestors who ventured onto the dry, dangerous plains. As grasslands spread, hunting and eating meat proved advantageous. The control of fire staved off predators. With the expansion of early humans into Asia and Europe, the challenges of the Ice Age helped hone the capacity for social cooperation. Speaking to one another helped pass on the traditions of toolmaking. Eventually, language allowed technological innovations to catch on and creative endeavors, including art, to blossom. All the way back to the earliest bipedal predecessors, one survival skill led to another, which spurred on still newer adaptations—all against the backdrop of the arid equatorial savanna and the frigid northern landscape.

This notion of a straight line trending toward a drier, colder climate has now been replaced, however, by evidence that past climate fluctuation looked more like a rambunctious zigzag. The survival conditions of human evolution were continually revised as climate oscillated between arid and moist and between cold and warm. Scientists now speak of the *environments* of human evolution, with emphasis on variability. And they consider the instability of environments, not merely the expansion of grasslands or glaciers, as having shaped the evolved characteristics of human beings.

This relatively new theme in the story of human origins is still a matter of hypothesis—an overall explanation that is tested again and again as new details come to light. One of the exciting challenges in the field of paleoanthropology is to better understand how our own ancestors may have evolved adaptations to *change*—to what researchers call *environmental dynamics*—rather than to any single setting or environmental trend.

Evidence about past climates is, naturally, an inspiration for this environmental variability hypothesis. An important record of global climate change comes from oxygen measurements in ocean microorganisms called foraminifera. Forams, as they are also known, use oxygen in their immediate surroundings to build their tiny calcareous skeletons. As the planet has cooled and warmed, the levels of different forms, or isotopes, of oxygen have changed. Because the lighter isotope more easily evaporates than the heavier one, it was extracted from the oceans and incorporated into the expanding ice sheets on land. When these glaciers melted, the water composed of lighter oxygen flowed back into the ocean. Oxygen measurements of the forams obtained from the deep-ocean floor document these fluctuations in Earth's temperature and ice volume over time.

How do we know about climate history?

Researchers have many different methods to interpret Earth's climate history.

Drilling sediment layers from lake and ocean floors produces long cylinders called cores. Drill cores are also collected from the thick ice layers of glaciers and ice sheets. The depths (or lengths) of cores correspond to time, with the most recent layers near the top. By studying trapped gas, pollen, charcoal, and microorganisms deposited in each layer, scientists can infer changes in past climate.

Tree rings and coral are proxies for climate because their periodic growth rates are affected by environmental conditions.

Cave formations provide useful information because they accumulate more or less rapidly in relation to how wet or arid, respectively, the environment is.

Fossil plants and animals are used to reconstruct ancient environments and climates. Fossilized leaves, seeds, bark, roots, and pollen are also direct indicators of ancient climate. Drawing upon the habitat preferences of living animals may help us use animal fossils to infer past habitats.

When we look at the oxygen climate curve all the way back to 70 million years ago, we see that Earth has indeed cooled dramatically, especially over the past several million years. Important details come into focus when one examines the past ten million years. This shorter record, which includes the era of human evolution, shows that the cooling actually involved sharp fluctuations between warm and cool. The oscillations began to pick up around six million years ago, near the time the earliest human ancestors originated. The genus *Homo* evolved later, during a time of even greater climate fluctuation. And the immediate predecessors of our own species, *Homo sapiens,* evolved as climate instability maxed out with the widest oscillations.

An equally compelling climate record focuses on the moist-arid fluctuations in Africa. Cores drilled from the bottom of the Mediterranean Sea give us a long archive of black mud layers that alternate with lighter bands of silt. The two types of sediment are derived from the Nile River. The dark layers are evidence of strong monsoons that washed black mud, rich with organic material, from the huge Nile watershed into the Mediterranean, while the light bands indicate

Sediments at Olorgesailie, in Kenya, contain a wealth of information about climate change over the past 1.2 million years. White and beige sediments reflect the area's alternation between a large lake and dry land, respectively. Brown sediments represent deposits of a river that flowed through valleys eroded into the older lake beds.

Left humerus

Right humerus

Homo neanderthalensis, *Shanidar 1*
About 45,000 to 35,000 years old
This Neanderthal survived a crushing blow
to the left side of his head and had
a withered right arm (left).

Australopithecus africanus,
Taung Child
About 2.8 million years old
A large bird of prey, probably
an eagle, carried this child away.

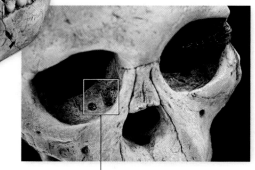

Eagle talon marks

arid times when rainfall declined and vegetation shrank. The record from the past five million years shows that early human species experienced periods of relatively stable climate interrupted by even longer periods when strong shifts between arid and moist conditions took place. Each interval lasted from a few thousand to several hundreds of thousands of years. This long-term switching back and forth between high and low climate variability had a persistent influence on the survival and ways of life of early humans as they evolved in Africa.

Environmental records from all over the world confirm the magnitude of climate fluctuation. The longest environmental archive from any continent comes from the Loess Plateau of central China, a 400,000-square-kilometer (150,000-square-mile) region that has accumulated windblown dust over the past 21.5 million years. There, researchers have documented continuous shifting between arid periods, when loess blew in from the northern deserts, and moist times when rich vegetation converted the loess into soil. Each of these phases lasted as long as tens of thousands of years. Although the fluctuations have occurred over many millions of years, they were especially large during the past 2.6 million years, around the same time as the glacial cycles of the Northern Hemisphere began.

Because instability in weather, food, and water inevitably posed challenges to survival, climate variability may ultimately help us make sense of the demise of species in our own family tree. Usually it's very difficult to pinpoint a single cause of extinction for any fossil species. Yet the habits of the large-toothed *Paranthropus boisei,* which lasted in Africa for a million years, or of the cold-adapted *Homo neanderthalensis,* which thrived in Eurasia for 200,000, were repeatedly tested as favored habitats expanded and contracted, sometimes more severely than others. Climate fluctuation also meant variation in the parasites, predators, and other dangers that challenged human ancestors where they lived. Over the past three million years in particular, powerful climate swings would have led to large fluctuations in supplies of crucial resources, contributing to occasional crashes in population size. All of these factors can influence the survival or extinction of species.

This leads to a curious finding in the study of human origins: Our closest evolutionary cousins—species that also walked upright, made tools, and had large brains—went extinct, even though these basic characteristics were at one time considered to be the hallmarks of evolutionary success in human beings. This finding brings us back to the question of adaptability.

CLIMATE CHANGE AND EVOLUTION

The rapidly expanding data on past climate change have started to recast our ideas about the evolution of human adaptations. What benefits did those adaptations offer as our ancestors confronted shifting conditions? The advent of upright walking, for instance, did not mean that our oldest ancestors abandoned the trees entirely. Instead, they walked across open terrain *and* climbed trees in more wooded areas. But later, as African environments varied dramatically between moist and dry, the ability to walk long distances would pay off well in the diverse landscapes encountered by *Homo erectus*. Similarly, at the dawn of stone technology, the basic toolkit—including hammerstones that could crush as forcefully as an elephant's molar and sharp-edged flakes that could cut as finely as a carnivore's tooth—would enable the earliest toolmakers of the genus *Homo* and possibly late *Australopithecus* to eat new kinds of food as conditions changed. Later still, as conditions continued to change, our evolving brain began to deal with richer and more complex surroundings and social interactions. Any improvement in how quickly brains could process information, call up memories, and forge new thoughts could have made the difference between survival and extinction.

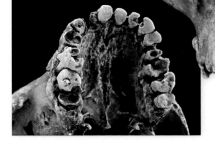

One of the most impressive and unusual aspects of humans today is the way in which we alter our surroundings. Creating stone implements, controlling fire, building shelters, growing and storing food: All represent ways of altering the immediate surroundings. Each made life a bit more predictable, furthering survival in surroundings that were prone to change. The entire package proved so successful that, eventually, the sole surviving hominin—*Homo sapiens*—was able to spread around the globe.

Certain abilities that evolved in earlier human species proved especially beneficial in times of change. Whether or not this variability hypothesis of human evolution stands up to all the tests of scientific data down the road, the drama of environmental change is now understood as a prominent backdrop to the story of our origins. In the end, it may offer new insights into the origin and current status of our own species.

Homo heidelbergensis,
Kabwe 1, "Rhodesian Man"
About 300,000 to 125,000 years old
This **Homo heidelbergensis** *lived with severe cavities (left) and perhaps died from a bone infection.*

FAQ:
How does evolution occur?

Genetic variation is fundamental to evolution. A population's gene pool undergoes slight changes every generation because of mutation and the recombining of parents' DNA in their offspring.

To survive, living things adapt to their surroundings. Natural selection provides an important mechanism for change in a population's gene pool over time. A genetic variation occasionally gives a member of a species an edge. That individual passes the beneficial gene on to his or her descendants. More individuals with the new trait survive and pass it on to their descendants. If many beneficial traits arise over time, a new species—better equipped to meet the challenges of its environment—can evolve.

DANGEROUS TIMES AT SWARTKRANS

In 1948, South African scientists began excavating Swartkrans, a remarkable cave site in the dolomite hills of South Africa. As they worked, they saw that the tawny sediments were filled with thousands of fossil animal bones. Among the bones were the remains of an early human species known as *Paranthropus robustus*.

More finds helped draw a picture of the way these early humans lived. The scientists discovered bone tools, used by *Paranthropus* to dig into termite mounds. They also found fossil animals, including antelope, zebra, wildebeest, and baboon—species whose presence indicated that the area, now scrubby grassland, had once been a mosaic of grasslands and woodlands.

One reason why bones accumulated at Swartkrans became clear after a young hominin skull with two circular punctures in the back was found. These holes match the spacing of the canines of a fossil leopard jaw from Swartkrans.

Was the young *Paranthropus robustus* the victim of a leopard attack? It's likely. The leopard might have ambushed a hominin digging for termites, made its kill, and then dragged its prey into a tree to prevent it from being stolen by a larger carnivore. Trees commonly grow near the mouths of underground caves like Swartkrans, so the skull and other bones would eventually have dropped down into the cave. The same thing probably happened to many other animals over the years.

Early humans, including this **Paranthropus robustus** *youth, sometimes fell prey to the dangers of their surroundings.*

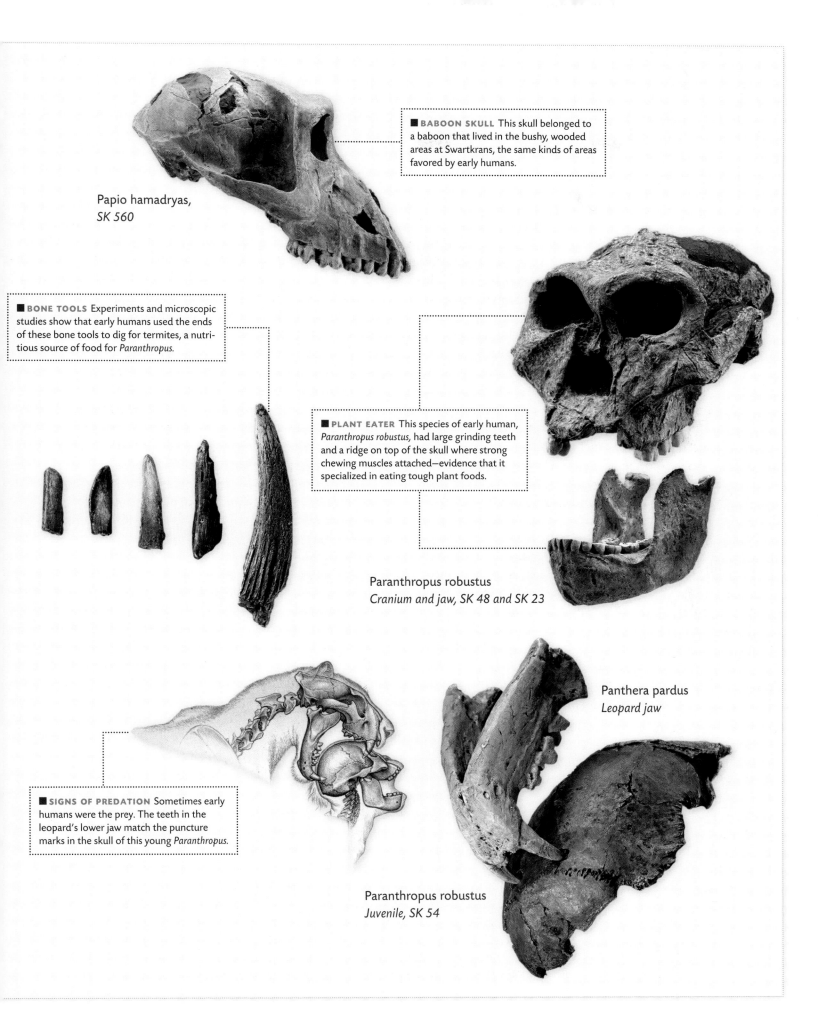

■ **BABOON SKULL** This skull belonged to a baboon that lived in the bushy, wooded areas at Swartkrans, the same kinds of areas favored by early humans.

Papio hamadryas,
SK 560

■ **BONE TOOLS** Experiments and microscopic studies show that early humans used the ends of these bone tools to dig for termites, a nutritious source of food for *Paranthropus*.

■ **PLANT EATER** This species of early human, *Paranthropus robustus,* had large grinding teeth and a ridge on top of the skull where strong chewing muscles attached—evidence that it specialized in eating tough plant foods.

Paranthropus robustus
Cranium and jaw, SK 48 and SK 23

Panthera pardus
Leopard jaw

■ **SIGNS OF PREDATION** Sometimes early humans were the prey. The teeth in the leopard's lower jaw match the puncture marks in the skull of this young *Paranthropus*.

Paranthropus robustus
Juvenile, SK 54

CHAPTER 4

THE FIRST STEPS

NEXT TIME YOU STROLL BAREFOOT ALONG THE BEACH, TAKE A LOOK AT WHAT YOU'VE LEFT behind. Notice the imprint made by your big toe and how it's aligned with the others? The compact toes help propel you across the sand, one foot after the other. Can you also see the arch, which gives spring to your step? Then look at the trail you've made. It's a wonder we can balance so well, walking straight forward, as each leg in turn supports our full body weight. The trail you've made points to one of the first milestones in becoming human.

While searching the Hadar badlands of Ethiopia in 1974, a team led by American paleoanthropologist Donald Johanson made an astonishing discovery. Arm bones, leg bones, ribs, a pelvis, a lower jaw, hand and foot bones: These and other fragments made up a 3.18-million-year-old female skeleton. In the Amharic language of Ethiopia, this skeleton is called Dinkenesh, which means "you are beautiful." In English, she is known as "Lucy."

Her discovery triggered one of the most successful hunts for early human fossils in the history of human evolution research. After years of dedicated searching, Lucy's species, *Australopithecus afarensis,* is now known from hundreds of individuals of both sexes and all ages. This species is thus the best known of the hominins of that geologic era, the Pliocene. *Australopithecus afarensis,* known from a handful of East African sites, ranges from about 3.9 to 3 million years old, after which Lucy's lineage met its demise.

Lucy and her fossilized kin give us clues about our ancestral way of life—and the success of one of the first experiments in becoming human: the ability to walk upright on two legs.

Previous pages: *Bushmen traipse across a hot Namibian salt pan. Our unique physique, brain, and behavior distinguish us from all other primates.*

Opposite: *Walking on two legs is a uniquely human characteristic that arose early in our evolutionary history.*

HOW HUMANS WALK

Humans are the only primates that regularly use two legs to move around with a smooth, striding gait. The bipedal way of life comes into play when we shop, carry our kids, run, dance, play games, and engage in all sorts of jobs that require our ability to balance and stride on two legs. We are unique in our dependence on walking upright.

The roots of this unusual behavior stem from the varied and agile ways in which primates move. Although primates are known for leaping through the

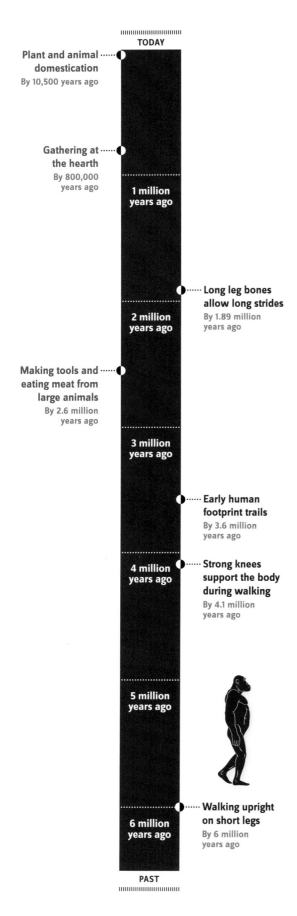

TODAY

Plant and animal
domestication
By 10,500 years ago

Gathering at
the hearth
By 800,000
years ago

1 million
years ago

Long leg bones
allow long strides
By 1.89 million
years ago

2 million
years ago

Making tools and
eating meat from
large animals
By 2.6 million
years ago

3 million
years ago

Early human
footprint trails
By 3.6 million
years ago

4 million
years ago

Strong knees
support the body
during walking
By 4.1 million
years ago

5 million
years ago

Walking upright
on short legs
By 6 million
years ago

6 million
years ago

PAST

trees and running on all fours, there are lemurs that hop across the ground on two legs; baboons that stand and totter short distances to pick blossoms from low-lying tree limbs; and all the species of apes that, on occasion, walk on two legs. Chimpanzees, gorillas, and orangutans walk bipedally by rotating their bodies to maintain their balance, shifting their weight over one leg and then the other. They do this with their hips and knees bent, unable to fully extend these joints.

Humans walk differently. We move along smoothly on two legs, with hardly any swaying from side to side or bobbing up and down. This approach to travel requires that the key parts work well together. The spine, hips, knees, and feet all have a say in the matter.

The short, broad shape of your pelvis rearranges how your muscles contract around the hip as your legs stride powerfully with each step. Because the upper pelvis is broad, rather than elongated as in apes, we have a large muscle—the gluteus maximus—that has become repositioned during the evolution of walking. This muscle gives everyone, like it or not, a big butt. The same muscle in a chimpanzee is positioned to the side. In fact, we're the only primate that has a backside of noticeable proportions. Walk behind a chimpanzee and you'll see there's not much there.

During walking, your knees flex, swing forward, and then naturally lock almost directly beneath the middle of your body, affording support at the center of gravity. Since the knee is at the midline, the femur must angle upward and outward toward your hip. An ape's body, by contrast, is supported on a nearly vertical femur. This means that the body sways, first over one knee and then the other as an ape navigates on two legs. At the far end of the leg, human and ape feet are wonderfully different. In us, the big toe is wide and stumpy, snuggled up to the abbreviated toes to help form a powerful yet resilient paddle. The energy-conserving arch is in the middle, and the ankle, heel, and other foot bones interlock to offer stability and support. Ape feet are more like hands, with elongated toes, the big one more like a thumb; no arch; and an overall flexibility consistent with a talent for grasping—a signal of some degree of dependence on climbing, standing, and nesting in trees.

This seems like a lot of anatomy to think about. Yet it enables us to understand the fossil clues that indicate an upright body posture and the two-legged rebellion that defined the start of our lineage.

FOSSIL EVIDENCE

The farther back in time we go, the more illuminating the fossil evidence becomes of the transition to bipedal walking. The best picture available comes from the skeleton of *Ardipithecus ramidus*, a 4.4-million-year-old hominin found in Ethiopia's Afar Depression. The main opening at the base of the skull shows that the spine connected farther toward the center of the skull than it would in apes. This is also true of another early hominin, the seven-million- to six-million-year-old *Sahelanthropus* from Chad. This forward point of attachment meant that the head was easily stabilized when the body was in an upright position. No bones are preserved from the body of *Sahelanthropus,* but those of *Ardipithecus ramidus* show it walked upright on feet more rigid than those of living great apes. When walking upright, *Ardipithecus* was supported by muscles that attached to a humanlike upper half of the pelvis. The lower half, though, was apelike and accommodated large muscles for climbing. A grasping big toe and massive apelike arms and hands confirm that *Ardipithecus* still felt at home in the trees.

We have the remains of *Orrorin*, although less complete, about six million years old. The most provocative part of its body is a fragmentary thighbone, or femur. The upper area, which acts like a bridge between the rest of the femur and the hip socket, shows a slightly thickened base compared with this area in apes. This small adjustment indicates that *Orrorin*'s femur could withstand the repeated stress of placing the body weight on each leg during upright walking.

Footsteps of human ancestors who walked upright 3.6 million years ago remain sealed in time by volcanic ashfall. Paleoanthropologist Mary Leakey, shown here at the site, discovered them at Laetoli in Tanzania in 1976.

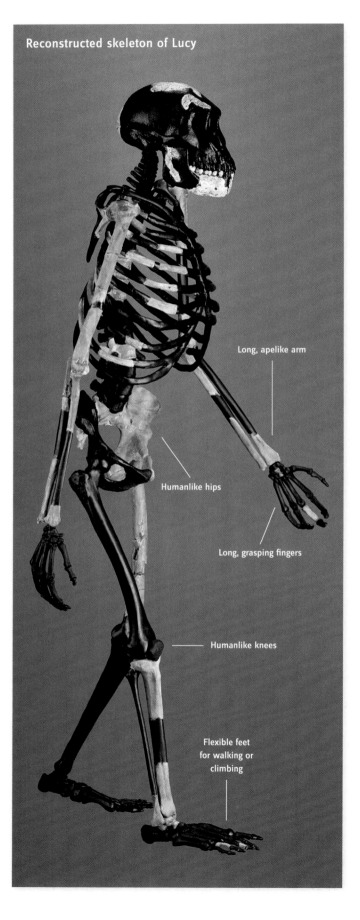

Reconstructed skeleton of Lucy

Long, apelike arm

Humanlike hips

Long, grasping fingers

Humanlike knees

Flexible feet
for walking or
climbing

Definite evidence of bipedal walking as a way of life shows up around four million years ago, based on a broadened knee joint. Again, a comparison of anatomy is helpful. The knee of a great ape is lightly built. This makes it hard for an ape to support its full weight on each leg, one at a time, over long distances. Although many of us are acutely aware of the wear and tear on our own leg joints, humans all have significantly broader, stronger knees that can more easily accommodate constant two-legged activity. A tibia of the species *Australopithecus anamensis,* dated about 4.1 million years old, shows the broadened expanse of the knee joint. This find suggests that this species had evolved the crucial adaptation at the knee for walking upright.

Fossil vertebrae dated more than a million years later provide even more evidence of the bipedal adaptations of *Australopithecus.* The spine of a chimpanzee or gorilla, as in other primates, makes a gentle arch from the pelvis to the back of the head. A human spine typically has a double curvature—the forward curve of the lower back, the backward arch of the area where the ribs attach. This double curve results from the slight wedge shape of the lumbar vertebrae and the soft intervening disks. This lower curve absorbs the small shocks generated by every footfall. Fossil vertebrae of *Australopithecus africanus,* dated around 2.5 million years old, show this important adaptation for upright walking in the lumbar region.

Step by step through the fragments of fossil evidence, we can thus see the anatomical rearrangements that are at the foundation of the human experiment in walking upright. These changes emerged in the hominins between 6 million and 2.5 million years ago, indicating that early humans who lived during this lengthy period combined apelike and humanlike ways of moving.

The Lucy skeleton, because of its completeness, presents the most striking union of traits. Coupled with evidence from other remains of this species, the skeleton of *Australopithecus afarensis* displays a humanlike pelvis and a small femur that was angled at the knee joint. These are solid clues to a life of walking on relatively short legs. Yet these traits were combined with long, powerful arms, curved finger bones, and longish toe bones—indicating a lifelong familiarity with climbing and grasping branches, and a foot more agile than our own. This amalgamation matches what we can expect when the course of evolution is caught in stop-action, crystallized in fossil remains, during the transition from one major way of life to another.

EARLY FOOTPRINTS

While Lucy deserves our attention for what a single fossil skeleton can show, the most enthralling single site relevant to the bipedal life of early humans is Laetoli, Tanzania. It preserves an astonishing series of footprint trails that are 3.6 million years old. As the oldest known tracks of early humans, the Laetoli footprints are as evocative of the human evolutionary journey as the footprints that our own species has left on the moon.

Embedded in an ancient landscape of hard volcanic ash, the Laetoli prints show where three early humans walked upright and straight ahead on two legs. The stride length of these individuals varied from 39 to 48 centimeters (15 to 19 inches)—evidence that these bipeds walked on shorter legs than most hominin species that evolved later.

In the same general vicinity, dozens of fossilized teeth and jaws of *Australopithecus afarensis*, Lucy's species, have been found. And they are so similar to the fossils from Hadar, Ethiopia, that an upper jaw from one site and a lower jaw from the other fit together as though they belonged to close relatives. Researchers cautiously ascribe the Laetoli footprint trails to Lucy's species.

The geologic layer that preserves the footprints also preserves equally remarkable evidence of the 3.6-million-year-old ecosystem. There are footprints of antelopes, an African hare, and a three-toed horse, among some 20 different species of fossil mammals. There are also prints of fossil birds, evidence of wasp cocoons, and even a report on the weather conditions—fossilized raindrops—at the time the footprints were made.

Inspection of all the clues allows us to reconstruct the scene around 3.6 million years ago. One in a series of eruptions of a nearby volcano deposited a layer of fine ash, 15 centimeters (6 inches) thick, on the Laetoli landscape. The ash is a type called carbonatite, which hardens when moistened by rain. Then, three bipedal humans walked across this plain, one of them carefully following in the footsteps of another. A later ashfall then sealed the footprints, which were buried until uplift of the region caused by earthquakes led to erosion and the natural exposure of the layers.

One can't help but wonder where these early humans were going or why they moved across this patch of ground as ash and rain fell from the sky. The image of one individual placing his or her feet in the footprints of another seems to betoken a familiar gesture of human intent—but we'll never know these early humans' motivation. And still, a larger question remains before us: Why did our earliest ancestors walk upright in the first place?

WHY WALK UPRIGHT?

Scientists have offered many creative reasons to explain why bipedal walking evolved early in the human journey. The most feasible possibilities take into

Opposite: The 3.2-million-year-old skeleton of Lucy shows her species, **Australopithecus afarensis,** *walked upright but was also still accustomed to climbing trees. The brown bones represent the parts of Lucy's skeleton that were excavated. The black bones were filled in using fossils from Lucy's species and knowledge of anatomy.*

Born to cling

For the first six months of life, modern human babies instinctively form a tight grasp around fingers, hands, or other objects that touch their palms. This is known as the palmar grasp reflex, and many infants are able to support their own body weight with this grip. Since this reflex is common to all primate infants, it's likely that all early human infants had it.

Millions of years ago, this reflex may have helped early human infants cling to their mothers, particularly during breast-feeding (when the reflex is strongest). But this may not be very effective for modern babies, since human mothers do not have long fur, and bipedal babies do not have grasping toes to make an effective foot grab.

While other great ape babies can hitch rides by clinging to their mothers' fur with all fours, human babies depend on their mothers to carry them.

account the fossil evidence and the record of environmental conditions at the critical time. They also consider the survival benefits and the costs associated with this significant evolutionary transition. The key is to figure out how walking upright improved the chances for survival and reproduction in the time and place where the transition occurred—beginning by about six million years ago in Africa.

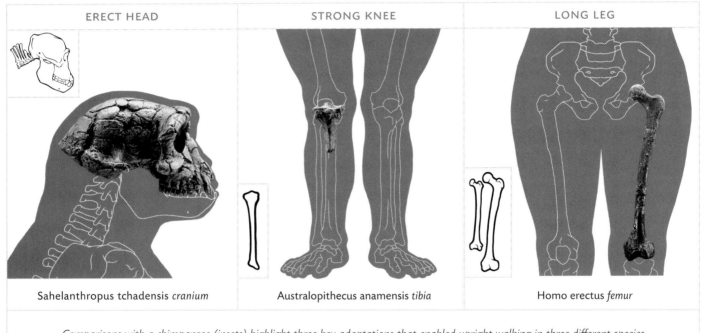

ERECT HEAD	STRONG KNEE	LONG LEG
Sahelanthropus tchadensis *cranium*	Australopithecus anamensis *tibia*	Homo erectus *femur*

Comparisons with a chimpanzee (insets) highlight three key adaptations that enabled upright walking in three different species of the human lineage. These traits are important for having balance, supporting upper body weight, and taking long strides.

How individuals use their energy while seeking food is an important factor to consider. One idea is that walking on two legs offered an advantage whenever the earliest hominins foraged for fruits, pods, or blossoms that could be picked from low-lying tree branches. Instead of repeatedly rising up on two legs and then returning to all fours, it could have been more efficient in terms of energy—and advantageous from an evolutionary standpoint—to keep going on two legs. We can't be certain whether the oldest bipeds sought food from tree branches nearest the ground. But we do know that chimpanzees and other primates practice bipedal behavior in situations where low-lying foods can be picked as they walk short distances.

How the muscles contract from the hip to the foot has a lot to do with the efficiency of moving on two legs. Measuring the amount of oxygen consumed while walking is one way to calculate this effect. Experimental measurements in primates show that bipedal walking on the ground is more energy efficient over long distances than four-legged travel. This suggests that any small change that could improve walking on two legs might well have increased the range over which our ancestors could search for food. It turns out that modern human bipedal walking is 40 to 50 percent more efficient than the four-limbed walk of a chimpanzee. The distances typically traveled by the two species seem to relate to the difference in locomotion.

But according to another scenario, the advantage of walking upright was in freeing the hands for carrying food, tools, or babies. A less traditional idea is that walking made early humans appear larger and more intimidating—a possibility that might have benefited the more aggressive members of a group.

A different approach to the problem comes from studying the environments in which the earliest bipeds evolved. Here we have a factor that would have influenced the survival strategies of all members of a population or species. One big question currently being debated is whether the earliest hominins preferred a single type of habitat or were adapted to a changing variety of settings. Some early human sites from six million to four million years old preserve seeds and fossil animal species that point to a moist and wooded setting. Other sites strongly suggest a mosaic of grassy, bushy, and forested surroundings. Stepping back, we can also recall that starting around six million years ago, the oxygen record from the deep sea suggests that global climate became more varied, with greater highs and lows in temperature and moisture.

There is one place where scientists have found early fossil humans in one layer after another during a lengthy era of climate change: Hadar. From fossil pollen grains richly preserved in sediments 3.4 to 3 million years old, scientists have identified 51 types of grasses, shrubs, and trees that grew at the Ethiopian site at different times. Along with fossil mammal bones, the pollen shows that the vegetation and climate oscillated repeatedly between cool and warm, wet and dry. The bones of *Australopithecus afarensis* are found throughout these deposits, meaning that this species was able to cope as the weather changed.

As the vegetation also shifted between forest and grass, with diverse bushes and shrubs always in the background, could it be that the bipedal behavior of Lucy's species proved very effective in adjusting to such change? One of the challenges of understanding the evolution of walking upright is discovering why moving on two legs *and* having a facility for climbing trees were combined for about four million years. At Hadar we have a clue that by the time of *Australopithecus* the two-legged approach was helpful in coping with change itself, perhaps by enabling hominins to traverse open spaces during the dry times to take advantage of the food and safety of the trees some distance away. The flexibility afforded by both walking and climbing may have enabled Lucy's species, and even earlier ancestors, to survive as the environment changed repeatedly.

Scientific evidence usually doesn't prove beyond doubt only one hypothesis when many plausible explanations exist. But only the most solid interpretations urge the discovery of new evidence about ancient environments, the biology of bipedal walking, and the factors involved in understanding these first steps in becoming human.

The flexibility afforded by both walking and climbing may have enabled Lucy's species, and even earlier ancestors, to survive as the environment changed repeatedly.

A troop of muscular **Australopithecus afarensis** *seeks food in the open woodland environment at Hadar. These hominins were probably as comfortable foraging in trees as they were on the ground.*

WALKING FASTER, FARTHER, MORE EASILY

It's not until around two million years ago that we find evidence of the next milestone in the evolution of walking. Fossilized bones from this time offer hints of a major change in the proportions of the body. In at least one lineage, the legs became elongated and the pelvis became larger and more robust. Walking long distances and running with endurance were the likely results of these changes. The evidence suggests a complete commitment to a rugged life lived on the ground.

Based on fossil bones found at the site of Bouri, in the Middle Awash region of Ethiopia, it appears that elongation of the legs had started to occur as far back as 2.5 million years ago. These bones are associated with an intriguing if fragmentary cranium that has been named *Australopithecus garhi.* Some scientists think that the skull may actually represent the large-toothed *Paranthropus.* The discovery, nonetheless, shows a peculiar combination of features that has the jury still out as to whether it represents its own unique lineage. If the leg bones do belong to the same species as the skull, it could mean that elongation of the legs—while the long, powerful arms of earlier hominins remained—occurred right at the cusp of the transition between *Australopithecus* and the earliest members of our own genus, *Homo,* around 2.5 million years ago.

By 1.9 million to 1.7 million years ago, we see the distinctive proportions of a body much like our own, with longer legs and shorter arms relative to the size of the torso. The earliest known species to exhibit this new look was *Homo erectus* in Africa.

What were the evolutionary advantages of this change? As any adult who has walked some distance with a young child can attest, longer legs allow one to go faster and farther with less effort. The same principle applies to this important milestone in human evolution. As experiments in energy output and anatomical studies suggest, bipeds that had shorter legs relative to the rest of their body expended more energy when they walked. Genetic variations that lengthened the legs conferred a longer stride, which helped early humans cover wide, open landscapes quickly and efficiently. In fact, environmental data from East Turkana in northern Kenya, and from Olduvai Gorge, Tanzania, indicate that East African environments fluctuated widely between moist and dry between 2.0 million and 1.7 million years ago. It is during this time that open grasslands began to spread. The ability to walk long distance with relative ease was a big advantage during this time.

As any adult who has walked some distance with a young child can attest, longer legs allow one to go faster and farther with less effort.

EVOLUTIONARY COSTS

The bipedal way of life has had many successes. The early lineage of *Ardipithecus,* the large-jawed *Paranthropus boisei,* and the muscular *Homo heidelbergensis* all were bipeds that thrived for hundreds of thousands of years. Our own species has dispersed to every corner of the globe by traveling on two legs.

Yet for all the advantages that walking presented to our early ancestors, this primal adaptation also has had its drawbacks. Distributing the full weight of the body on just two limbs can cause painful consequences ranging from lower back pain and slipped disks to arthritis of the knees and fallen arches in the feet. Walking upright also eventually limited the size of the birth canal because the hip joints could be only so far apart without compromising efficient two-legged striding. This ancient constraint led to the difficulties of giving birth to large-brained babies.

Walking upright laid the foundation for all that came later in our evolutionary history. But sometimes it's the painful consequences we feel that serve as a personal reminder of this oldest event in our origin.

FAQ:
What can we learn from ancient footprints?

Fossil hominin footprints and trackways are traces of ancient behavior. Through them, scientists can infer how early humans walked, what they were doing when the tracks were made, and the environments they lived in.

Scientists uncovered the earliest known hominin tracks, made about 3.6 million years ago, at Laetoli, Tanzania. These prints represent bipedal walking, but debate continues about whether these tracks represent a fully modern human gait.

Footprints found recently near Lake Turkana in Kenya more closely resemble modern human footprints. These 1.5-million-year-old tracks show deep imprints made by the heel and the ball of the foot during each stride—hallmarks of the modern human gait.

CHAPTER 5

FAMILIES AND GROWING UP

CARING IS FUNDAMENTAL TO WHO WE ARE. WE DON'T GIVE CARE ALL THE TIME, OR TO A consistent degree. Yet being cared for is essential to growing up. We are the only species with a childhood, reliant on parents and others for the essentials of life well after we have stopped nursing. Throughout our lives we share food with others, pool our resources, and rely on economic ties. What is the origin of togetherness and the unusual ways humans show it?

Opposite: *Newlyweds in Provence, France, celebrate with family and friends. The pairing of adults, and the support of a network of kin and others, have very ancient roots.*

The social life of early humans would seem difficult to figure out. It's hard to imagine how dusty artifacts or bones could indicate how fast our ancestors grew up, or how adult males and females interacted so long ago. Yet careful study of fossil bones and archeological clues provides surprising evidence of when certain unique qualities of human social life emerged.

All of the great apes, and virtually all primate species, live in social groups made up of individuals that interact in complex ways. Primate species grow up slowly compared with other mammals, and their young engage both in rambunctious play and attentive learning. Primate adults protect their young and have intricate relationships with other adults that are often played out in the nuances of grooming, such as how long one individual inspects and picks through the fur of another. In all great ape species, adults form long-term emotional attachments, concern themselves with status, and pay a lot of attention to sex. They show their canines and display other facial expressions in bluffing and friendship; they can punish and even kill. Given these basic elements of our primate heritage, it is virtually certain that our ancestors also engaged in a wide range of social interactions.

The evolution of social life over the past six million years is central to human origins. Consider how we differ from the other great apes: Humans today have an extended period of growing up, about six years longer than chimpanzees. The attention and energy that parents and caregivers devote to offspring is beyond anything seen in our primate relatives. Almost all humans focus the care of young and infirm individuals at a home base—a safe, special location where family and other group members can be expected to return daily. Adult males and females also create strong economic bonds based on sharing resources and mutual dependence. Human mating patterns are remarkably varied: from one lifetime pairing to a series of pair-bonds; one male may wed multiple females, and one female wed multiple males. Virtually the entire range of mating patterns found in other mammals occurs in human adults.

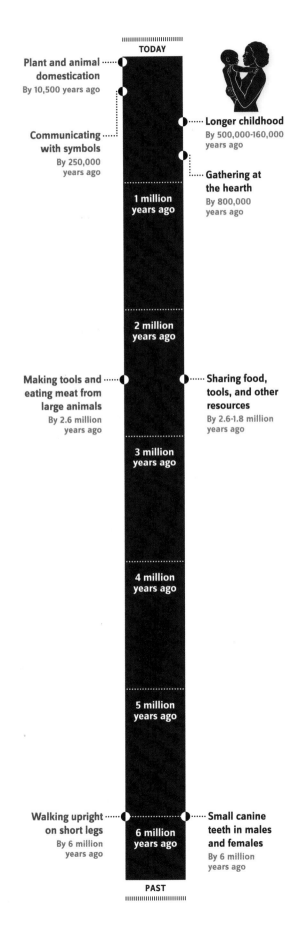

TODAY

Plant and animal domestication
By 10,500 years ago

Longer childhood
By 500,000-160,000 years ago

Communicating with symbols
By 250,000 years ago

Gathering at the hearth
By 800,000 years ago

1 million years ago

2 million years ago

Making tools and eating meat from large animals
By 2.6 million years ago

Sharing food, tools, and other resources
By 2.6-1.8 million years ago

3 million years ago

4 million years ago

5 million years ago

Walking upright on short legs
By 6 million years ago

6 million years ago

Small canine teeth in males and females
By 6 million years ago

PAST

The popularity of the Internet and online social networking are but an extension of two other peculiar qualities of human social life. First, virtually all of us belong to, and identify with, a broad community of people, many of whom we have never met face-to-face. We see ourselves as members of nations, professional societies, or other groups that are defined symbolically. Second, we all live in communities that develop strong social ties and dependencies not only with neighboring groups but also with people far away who are unfamiliar. We may donate our time or funds to disaster victims in other countries or join with those who speak a different language to fight a common enemy. The social networks on which these human activities are based are unknown in the lives of our primate kin.

The milestones in our evolving social life are among the most fascinating topics in the study of human origins. When did human social life begin to change? And how can we tell?

SMALL CANINE TEETH

The fossil record offers some tantalizing evidence about early human social groups. The discovery of remains of at least 17 *Australopithecus afarensis* individuals all together at Hadar, Ethiopia, in a layer about 3.2 million years old—sometimes called the first family site—shows that these individuals lived together as social creatures. At Laetoli, Tanzania, where the footprints of three *A. afarensis* individuals were found, one set shows a slight distinction between the right and left sides, suggesting that this individual might have been a female carrying an infant on her hip. The other set of prints could well have been made by a larger adult male, and the third individual had carefully stepped in those footprints.

One of the first hallmarks of emerging humanity, small canine teeth, are found even before Lucy's time, in the earliest members of our lineage. In most other primates, males possess large canine teeth to threaten, bluff, and bite others—characteristics of male social interactions. Honing of their sharp back edges against the lower premolars maintains their pointed, daggerlike look.

In the oldest known hominin species, the male canine teeth had become smaller, barely projecting beyond the level of the other teeth. The upper and lower jaws of *Sahelanthropus tchadensis,* seven million to six million years old, already exhibit smaller canine teeth than those seen in any of the great apes; the shape of the premolars had changed such that sharpening of the upper canine would have been impossible. Even more impressive evidence comes from *Ardipithecus ramidus,* by 4.4 million years ago. Since many canine teeth of this species have been found, it's certain that both males and females are represented in the sample. Yet the largest ones are not much bigger than the smallest, and they all show the roughly diamond shape typical of our own canine teeth. Male canines had become "feminized."

The disappearance of the main anatomical weapon that primates used to intimidate one another speaks to a pivotal change in social life near the start of the human lineage. Males no longer possessed even the ability to threaten one another with a quick gape of the mouth. Given the importance of the canines in male dominance displays and competition, the earliest hominin males must have been rewarded in some way for having less threatening grins, either by cooperative ties with one another or with females.

Clues to the female side of the story have yet to be found in the oldest hominin fossil record. But there are hints in the remains of *Ardipithecus ramidus,*

especially in the skeleton known as Ardi. Although the slender build of certain parts of the skull suggests that the skeleton is female, its bones are about the same size as those of any other individual of her species, suggesting that males and females were about equal in size. The miniaturized male canines and a female body size nearly equal to that of males imply that males and females had begun interacting in friendlier, more cooperative ways than those of most other primates.

PRIMATE HERITAGE

The basic social bond in all species of primates is the one between the female and her infant. Males exhibit a wide variety of roles. But only in humans do we

The intimidating canine teeth of a chimpanzee are honed as they slide into gaps between teeth. Both the large canines and the gaps were lost as the human lineage evolved.

Following pages: *The social life of our closest cousins, the Neanderthals, may have strongly resembled our own. But whether Neanderthals communicated by speaking a complex language remains unknown.*

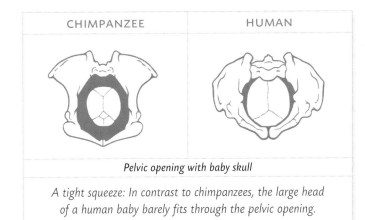

CHIMPANZEE	HUMAN

Pelvic opening with baby skull

A tight squeeze: In contrast to chimpanzees, the large head of a human baby barely fits through the pelvic opening.

Raising offspring involves long-term responsibility for human adults. Here, Australian women teach youngsters how to make a string turtle symbol.

see substantial male involvement in providing food, protection, and direct care of the young. Even in those few primates, such as long-armed gibbons, where females and males pair up and are nearly equal in size, males are largely occupied with patrolling territorial boundaries and fending off other males, thus maintaining the long-term pair-bond with the adult females in their territories. At some point during human evolution, males became much more deeply involved in the basic female-infant social unit. The reduction in canine size indicates that this transformation may have taken place near the origin of the human lineage.

Comparisons with chimpanzees provide some researchers with significant perspective on this evolutionary milestone. Female chimpanzees develop large, showy swellings on their rump when they are ovulating. These estrus swellings, which signal their readiness to conceive, mark the time when females are most likely to elicit male sexual attention. Human females have no such obvious displays, and so the timing of ovulation is essentially unknown to males. Human sexual activity is not confined to a few days cyclically associated with ovulation. According to one scenario, an early human male had little assurance that any given infant was his—that is, the result of his own mating activity. That could become clearer, though, for a male who was able to establish a special

bond and a daily role in provisioning a female and her offspring. Males who adopted this approach had greater reproductive success, and passed on their genes more often than those who didn't.

To other students of human evolution, however, this reasoning justifies one idealized version of family life. A broader initial comparison, for example, indicates that overt estrus swellings, as in chimpanzees and bonobos, are quite rare among primate species. Since the lack of a showy display is the primate norm, the absence of estrus behavior in humans might not even require any special explanation.

The more pertinent question is when and why adult males began to devote attention, resources, and care to the young. Only later in time, well after the earliest human ancestors lived, do clues about the development of this aspect of social life emerge.

THE ECONOMIC BOND

Humans are very peculiar in that when we search for and locate food, we usually don't eat it right away. Instead, we tote it elsewhere with the expectation of eating the food in the presence of others. People today take this activity for granted, yet it is an odd way to behave, given the tendency of all other mammals to "eat as they go." Foraging and eating with little delay go together among mammals. By contrast, the next time you go to the grocery store, stop to think about all the steps involved in growing, processing, transporting, and stocking the food you are purchasing, along with the fact that you usually don't start peeling, cutting, and opening packages right there in the aisles. We usually obtain food many hours—sometimes many days—before we eat it. Restaurants are about as close as we get to the "eat when you find it" approach all primates take.

This behavior of bringing food to someone else might lead us to think of birds bringing food to their young in the nest. Hyenas, wild dogs, and other carnivores likewise provision offspring stationed in the lair or den. But what humans do is to transport and share food in elaborate ways with other adults as well. This behavior carries immense expectations and obligations among the males and females who make up the household.

Human families are remarkably diverse, yet all societies recognize not only the reproductive bond between female and male but also the economic bond—the responsibilities of adult males and females to share and combine resources. Although in most cultures family life starts with the process of pair-bonding between adults, all societies recognize the vital importance of *paid-bonding*—a turn of phrase that emphasizes this economic relationship. In many traditional cultures this relationship is reflected in dowries and bride-prices—payments offered and expected in return for permission to marry and unite families. The financial aspects of divorce also echo the unique economic obligations that underlie the bonds between a male and female joined in a family. The merging of resources and energies is a centerpiece in the evolution of human family life.

Have the sexes always differed in size?

Many primate males are bigger than the females of their species. The degree of this difference corresponds to variations in species' mating practices. When males are considerably bigger, as in gorillas, it is common for a single dominant male to have reproductive access to several females. A smaller male-female size difference, as seen in gibbons, is often associated with long-term pair-bonding. According to some scientists, the males of primate species in which differences are large typically provide less attention to offspring.

The earliest hominin for which both male and female fossils are known is *Ardipithecus ramidus*, in which there may have been little male-female size difference. Yet males in some species of *Australopithecus* and *Paranthropus* were considerably larger than the females. Since the evolution of *Homo erectus*, there has been a trend toward diminishing size differences.

The milk teeth (yellow) and unerupted adult teeth (white) exposed in a computed tomography (CT) image of the skull of a three-year-old Australopithecus afarensis *reveal an apelike growth pattern. The pink area indicates the brain's outer surface.*

HEARTH AND HOME

The archeological record provides intriguing clues about when the delayed eating of food, its carrying, and its sharing first evolved. The development of campsites or home bases—places where adult males and females bring food and pool their resources—represents a vital sign of these unique qualities of human social life.

A step in the process is recorded in the earliest archeological evidence. At sites like Gona and Bouri in the Middle Awash of Ethiopia, humans 2.6 million to 2.5 million years ago transported stone tools from hundreds of meters to as much as a few kilometers from the original rock sources. Excavations at the site of Kanjera South in western Kenya show that, by two million years ago, meaty animal bones were transported repeatedly to the same place, and so were certain kinds of stone used in toolmaking from as far as 12 or 13 kilometers (7 or 8 miles) away.

Many of these ancient archeological sites were places that members of the social group visited repeatedly. Yet some sites were also visited by carnivores, which gnawed on parts of carcasses that were incompletely butchered. These places did not necessarily afford the safety of a home base where early humans would sleep, and young, old, and ill individuals could be cared for. Nevertheless, archeological sites between 2.6 million and 1.8 million years old preserve evidence that early toolmakers delayed their consumption of food and carried it over and over to places where they processed it and probably shared some morsels with other group members.

By 1.7 million years ago, when *Homo erectus* females evolved a larger body, the energy-intensive demands of maintaining the brain, the body, and a fast-growing hominin infant almost certainly strained a female's ability to cope. Helpers in nurturing the young, or what researchers call alloparents, would have proved essential. Adult males also would have been an important part of the mix. By caring and helping provide for the young of particular females, who also shared the foods they collected, adult males could make a decisive difference in the survival of their offspring.

FAQ:
Why do we play?

Play is not just fun; it is fundamental to the healthy development of brains and bodies. In humans and other primates, play helps establish bonds between adults and young, and its significant physiological benefits extend into adulthood. In addition to stimulating brain maturation, these include reduced stress, increased energy, and increased longevity. In humans, play allows children to develop imagination, dexterity, strength, and cognitive and emotional skills. A number of factors in modern life threaten play. Poverty, child labor, hurried lifestyles, and pressure to perform academically are among them. The United Nations High Commission for Human Rights has proclaimed play as a right for every child.

The size of the human brain is among the most important factors determining the amount of care newborns and children require. As we will explore in a later chapter, the most rapid rate of brain expansion in our evolutionary history was under way by 800,000 to 500,000 years ago. Babies were born with a large head relative to the size of the female birth canal by around this time. Parental care involving more than the mother was essential for the evolution of the human brain.

Pieces of burned flint (above), each about 1 centimeter long, indicate the presence of hearths and provide evidence for controlled use of fire, at the 790,000-year-old site of Gesher Benot Ya'aqov in Israel (artist's depiction, left).

The development of distinctively human home bases probably arose around this time, indicated by the building of hearths and shelters 800,000 to 400,000 years ago. Gathering places with these long-term structures signaled a definite departure from the nightly nesting sites quickly constructed by great apes, other primates, and probably the earliest human ancestors.

The evolutionary milestones that ultimately led to hearth, home, and a safe place for young to grow and play meant that females and males would come to rely on one another in the daily quest for food. This new type of relationship, which was the seed of the rituals and bonds of marriage across human societies today, greatly expanded the opportunities for food sharing and parental care in the immediate ancestors of our species.

A LONG TIME TO GROW

The most influential factor in human social life is that everyone is born incapable of surviving without years of dedicated nurturing. This begins with nursing by the mother and includes a protracted period when human parents continue to provide for their young after weaning. In all other primates, as soon as youngsters are weaned from the mother's milk, they are self-sufficient in foraging for themselves. Human children are capable of going out and finding food, but childhood is primarily defined by the continuing dependence on adults for care and feeding.

One of the defining features of humans is our prolonged series of life phases. In addition to childhood, humans are the only primates that have an even later phase of maturation—adolescence, the teenage years—characterized by a growth spurt and a delay of reproduction for several years after puberty.

Fossilized bones of immature individuals enable researchers to figure out how long it took our earlier ancestors to grow up. These discoveries include a skeleton from Dikika, Ethiopia, of a three-year-old *Australopithecus afarensis* individual who lived about 3.3 million years ago and two remarkably complete skeletons of two-year-old Neanderthal children discovered in Dederiyeh Cave, Syria, dated 70,000 to 50,000 years old. The eruption of teeth in the Dikika individual indicates an apelike rate of growth, although the brain may have developed at a slightly slower, more humanlike tempo. The Neanderthal children from Dederiyeh, by contrast, show patterns of bone, brain, and dental development similar to the more extended rate of modern human growth.

At 1.5 million years old, the "Turkana Boy" skeleton from Nariokotome, Kenya, represents a *Homo erectus* youth. Microscopic study of the enamel layers of the teeth, which are deposited on a daily basis, shows that this individual grew at an apelike rate. Although his second molars had already erupted, an event that typically occurs in our species at around age 11 or 12, the Turkana Boy was closer to 8 years old when he died. This implies that *Homo erectus* didn't have the distinctive childhood of modern humans.

The oldest example of a prolonged growth pattern typical of our species comes from a 160,000-year-old fossil jaw of *Homo sapiens,* unearthed from Jebel

Humans are unique among primates in having long, distinct periods of childhood and adolescence. These stages enable us to learn, play, socialize, and absorb important experiences before adulthood. Also unique among primates: the long lives we live after prime reproductive age.

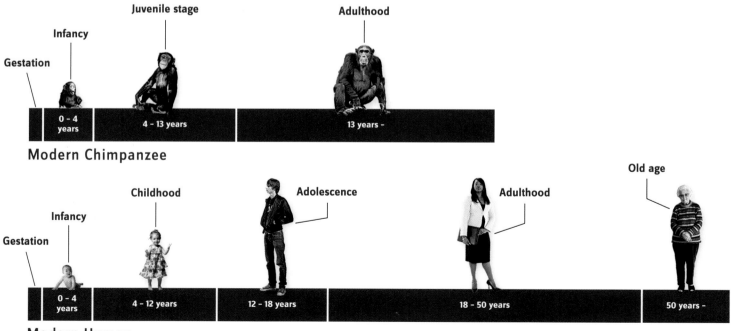

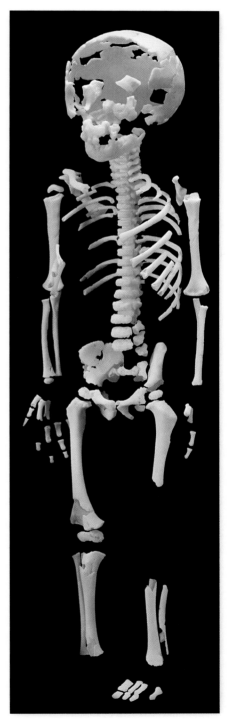

The 70,000- to 50,000-year-old skeleton of a two-year-old Neanderthal child (below), 83 centimeters tall, was found buried inside a Syrian cave at Dederiyeh (left). It grew at around the same rate as a modern **Homo sapiens** *child would.*

Irhoud, Morocco. Careful study of the enamel layers of this individual showed that they developed just like those of a seven- or eight-year-old modern human child, confirming the age of this individual at death and the presence of a prolonged childhood phase.

Two other unique qualities of human social life are important to note. First, we have the capacity for a more rapid rate of reproduction than our primate relatives do. The interval between births in many cultures today averages around four years, which means that a mother might have the responsibility for one or two children while she is pregnant or nursing a newborn. The potential for rapid reproduction, coupled with the prolonged maturation of the young, made families and social support networks critical to survival by the time *Homo sapiens* evolved.

The second point is that humans live a long time after the prime reproductive years. Our species has grandmothers and grandfathers. Elderly people possess an accumulated body of experience that assists in building knowledge of the world. Although the fossil record doesn't come with labels telling us who the earliest long-lived grandparents were, the remains of an elderly male discovered at Dmanisi in the Republic of Georgia, dated around 1.78 million years old, provides an interesting case. His upper and lower jaws show a great deal of bone loss; his teeth had fallen out well before he died. Since the lack of teeth would have proved fatal almost immediately, he was most likely fed by younger individuals in his social group, leading some researchers to suggest that this is the earliest known evidence of empathy. At least by this date, it was within the capacity of our ancestors to care for the elderly, and for newborns, children, and one another regardless of age. This elaboration of social life offered a strong safety net in a risky and challenging world.

CHAPTER 6

TOOLS AND FOOD

OUR EARLY ANCESTORS MADE IMPLEMENTS BY CHIPPING THE EDGES OF ROCKS. Flaked stone defined the moment when we first altered our surroundings. Sharp edges and weighty hammerstones were the first food processors, providing a way to cut and crush foods we could not otherwise chew. Piles of discarded artifacts were the first hint of someone who marked the landscape with trash. Carrying rocks from miles away, our ancestors searched far and wide for the resources of survival and comfort. The earliest toolmakers thus held in their hands a seed of no small importance to our lives today.

Around four million years ago, early humans began to experiment with processing new kinds of food. They did so by evolving larger teeth. The enlargement was focused on the back teeth: the molars and the premolars. Eating a varied diet was important, and this included grinding and crushing tough foods, with the help of the expanded chewing surfaces of the back teeth. The diversity of microscopic pits and scratches on the teeth of these *Australopithecus* ancestors confirms that they possessed a flexible diet. In these early human ancestors, any sustained change in the diet required an evolutionary change in the teeth.

Diet—the foods that an organism obtains and eats—is important because it's the source of energy and nutrition, which are essential to survival, reproduction, and an organism's ability to adjust to its surroundings. Anything new that influenced how our early ancestors crushed or sliced or otherwise made food edible was an important factor in human evolution.

Between 3 million and 2.5 million years ago, the australopiths continued to flourish in southern Africa, but elsewhere on the continent they became extinct. This was a period of strong climate change globally. It involved pulses of expansion and contraction of glaciers around the northern ice cap, greater aridity and grassland expansion in the low latitudes, and a widening range of fluctuation over time between cold and warm, dry and wet. By the end of this interval, two new lineages arose that defined the course of human evolution during the next 1.5 million years.

The *Paranthropus* lineage took food chewing to a new level. Molars and premolars expanded to enormous proportions; faces and mandibles enlarged to absorb the strain of heavy chewing; and the muscles of mastication created stresses

Opposite: *A butcher in Vietnam displays an array of meaty choices. Our interest in animal protein, and our ability to make tools to access it, go back at least 2.6 million years.*

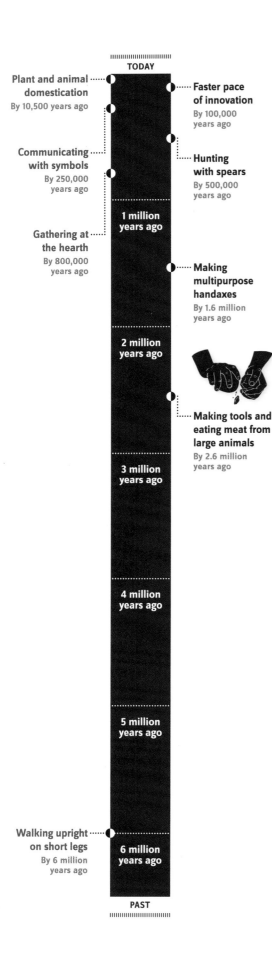

TODAY

Plant and animal domestication
By 10,500 years ago

Communicating with symbols
By 250,000 years ago

Gathering at the hearth
By 800,000 years ago

1 million years ago

2 million years ago

3 million years ago

4 million years ago

5 million years ago

Walking upright on short legs
By 6 million years ago

6 million years ago

PAST

Faster pace of innovation
By 100,000 years ago

Hunting with spears
By 500,000 years ago

Making multipurpose handaxes
By 1.6 million years ago

Making tools and eating meat from large animals
By 2.6 million years ago

sufficient to lead to the growth of a bony ridge at the top of the skull where the most prominent chewing muscle was anchored. This hyped-up approach to dental evolution was highly successful. The species that evolved the most dramatic expression of the chewing apparatus, *Paranthropus boisei,* thrived for more than one million years in East Africa.

This chapter, though, concerns the exploits of the other lineage—the one with smaller jaws that eventually relied on sharp flakes of stone and established the foundations of human technology. This is our lineage, the genus *Homo.*

"TEETH" MADE OF STONE

The earliest evidence of stone toolmaking is around 2.6 million years old, recorded in archeological digs at Gona in the Middle Awash Valley, Ethiopia. The tools consist of sharp flakes; stone cores, which have overlapping scars along the edges where flakes were struck; and hand-size hammerstones that were used to hit the cores. This flake-core-hammerstone toolkit is the most basic yet discovered, and is made of volcanic rocks that early humans found locally. Outcrops of hardened lava rock occur throughout the African Rift Valley, along with rarer sources of quartz and chert that also provided stone for the earliest toolmakers.

No evidence of stone tools is associated so far with hominin species older than 2.6 million years back to the origin of the human lineage. By 2.6 million years ago, however, some groups of early humans had enough experience with toolmaking to be quite adept. They had already mastered the mechanics of hammerstone-on-core percussion such that they were able to carefully and powerfully remove sequences of flakes from both sides of a stone cobble. They identified angles of less than 90 degrees, which are critical to the craft of stone flaking, and they could predict well enough how the sharp slivers would separate from the striking point to set up the next flake removal.

At the site of Lokalelei, west of Lake Turkana in Kenya, archeological sites 2.3 million years old have produced so much stone tool debris that the flakes and cores can be put back together—in a painstaking process called refitting, which is done after the excavated artifacts are brought back to the laboratory. The Lokalelei refits, which involve dozens of flakes, show that the toolmakers by this time could carefully conserve stone and plan to reduce a rock, one precise strike after another, and produce many sharp flakes from even a small amount of stone.

The skill level of these early toolmakers suggests that the craft of knapping stone actually started sometime before 2.6 million years ago, perhaps in dry stream channels where many cobbles and naturally broken stones may make it hard to recognize the initial attempts at stone knapping. The first recognizable stone tools, at 2.6 million years old, occur in dense concentrations; these archeological sites record the oldest stone tool *accumulations*—places on the African landscape where hominins repeatedly carried and used their basic toolkit.

Making and carrying stone tools helped early humans solve two major survival challenges. First, stone tools represented the first opportunity for food processing outside of the body itself. Sharp flakes and pounding implements improved significantly upon the teeth inside the mouth of any other hominin. A new universe of foods became available to exploit. Sharp flakes were used to cut tough animal hides, butcher meat from bones, and detach limbs from a carcass for transport elsewhere before dangerous carnivores approached. Stone edges could also whittle a pointed digging stick to uproot tubers or dig for pools of water underground. These activities were impossible using only teeth and hands inherited from earlier ancestors. Hammerstones proved useful not only in knapping sharp

Opposite: *Tools of our closest kin, the chimpanzee, serve varied purposes. Hammer and anvil are used to crack hard oil palm nuts (center); twigs (bottom) can be poked into mounds to capture ants; and a sharpened stick (top) can spear bushbabies while they sleep.*

Below: *Chimpanzees at Bossou in Guinea crack nuts while in a social group, which permits the knowledge of tool use to be passed from generation to generation.*

flakes but also in pounding nuts, crushing pods and tubers, and cracking bones to obtain the nutritious fatty marrow. The basic toolkit allowed the preparation and chewing of food to begin outside of the body and made essentially any item that could be chewed by other mammals accessible to the toolmakers.

The ability to carry the tools away from the outcrops or cobble sources solved a second challenge: It enabled these new "lithic teeth" to be taken to any food source. Early humans didn't have to find the food and then bring it to the near-

Hammerstone

Core

Flake

The first toolmakers used rounded hammerstones to strike sharp flakes from another rock, called a core—as shown in the illustration above.

est rock outcrop; instead, a portable core and hammerstone could be carried while foraging or left in places where other members of the social group might meet. The portability of stone tools was as critical as the technique of stone-on-stone percussion itself, making it possible to bring food and the newfound ways of food processing together in the same place. In this sense, transporting tools was a precondition for the profound changes that were to occur in human social behavior, from food sharing to the development of home bases.

Gaining access to new energy sources was a vital milestone in human evolution. As soon as stone tools appear, we also find animal bones with slicing marks and damage caused by butchery with stone tools. At the site of Kanjera South in Kenya, a microscopic study of two-million-year-old stone tools also identified the distinctive polish and abrasion on the tool edges that result from cutting up starchy tubers. The toolmakers could focus on the most energy-rich foods, which eventually helped fuel the enlargement of the brain.

It is less clear when, exactly, early human toolmakers became dependent on animals to eat, or when it became possible to hunt on a regular basis. Butchered remains of immature zebras and small gazelles, less than 50 kilograms (110 pounds) in size, are well represented at Kanjera South. Such animals were very likely captured by hand, since small prey are typically consumed immediately by African carnivores, leaving little or nothing for scavengers. There is nothing to suggest, however, that early humans had the tools to bring down larger animals like an adult zebra or wildebeest, much less a hippopotamus or elephant, even though the butchered remains of these creatures are found at archeological sites between 2.6 million and 1.5 million years old. The excavations do make one thing clear: Early human toolmakers were competent in acquiring meat and marrow from many different species, and they did this in a variety of woodland and grassland settings. This evidence underlines the opportunism that typified the earliest stone knappers.

USE TOOLS TO MAKE TOOLS

Making tools was once considered to be the defining characteristic of human beings. Yet ever since Jane Goodall first described the chimpanzees of Gombe Reserve, Tanzania, and how they carefully prepared sticks to capture termites, toolmaking has also become recognized as part of the repertoire of great apes. Chimpanzee tools range from leaf sponges that sop up water to walking sticks that help them wade through wetlands; from dipping sticks carefully inserted into ant nests to wooden branches used to break open termite mounds; from hammerstones and anvils for cracking nuts to sticks used as spears for stabbing small primates.

Chimpanzee tool use, toolmaking, grooming, and other activities are even described as cultures, which means that groups in different regions exhibit a unique set of behaviors. The use of hammers and anvil stones to crack nuts, observed in some West African groups of chimpanzees, is especially interesting because the stones are carried from one nut tree to another. Food debris—the nut shells—accumulates alongside the battered and pitted tools. The fact that it can take months or even years for an observant young chimpanzee to learn how to use stone hammers to crack nuts, or to prepare just the right kind of stick to "fish" for termites, offers further evidence that the capacities shared among the great apes and our common ancestors set the stage for stone toolmaking in early humans. There is little question that hominins inherited from earlier primates an ability to manipulate objects and alter them by hand to make useful tools.

Early human ancestors' toolmaking differed from chimpanzees' in three main ways. First, chimpanzees use their hands to modify sticks, leaves, and other natural objects, whereas the oldest known hominin technology involved the use of tools to make other tools. Wielding a hammerstone to strike flakes from a core characterized the oldest known stone toolmaking. Using tools to make tools remains a defining characteristic of early human technology.

Second, the earliest stone toolmakers created clusters of hundreds of flakes, cores, and hammerstones, which were associated with food remains transported to the sites. While chimpanzees do bring tools to nut trees and other food

Rock—or stone tool?

Stone tools reflect intentional modification by a human, not random battering by natural forces. Early stone tools were made by purposefully using a hammerstone to strike and break sharp slivers from a rock. The result was a pattern of overlapping scars along the rock's edge, numerous sharp stone flakes, and battered hammerstones. The presence of all these altered stones together helps archeologists identify stone tools. In later stone technology, the steps in the manufacturing process must be taken in a specific order so that the shapes of the tools are correctly crafted.

Context is also important, since it can be difficult to distinguish tools from naturally broken rocks at the bottom of rocky cliffs or where fast-moving water, as in rivers, can bring rocks together naturally. Stone tools are most easily recognized in places where rocks could not have been introduced by natural geologic forces.

sources, early humans carried both the tools and the food to other, specific places on the landscape, perhaps near water or under a shade tree. This "double transport" system involving both food and tools was an important innovation that ultimately transformed human social behavior.

Third, the earliest known stone tools significantly revamped the ecology of human ancestors, leading them to exploit large mammals that were rich in protein and fat. This was the only time in primate evolution that this major shift in diet occurred. Chimpanzees and a few other primates are known to kill and feed on small prey, but small prey make up no more than 2 to 5 percent of the diet. Making stone tools, by contrast, enabled early humans to feed on animals much larger than themselves and to take a giant step into the hazardous domain of the African predators.

At first, these differences were probably matters of degree. Yet the expansion of diet and lifestyle that carefully chipped stones offered eventually led humans to exploit a huge number of energy sources and opportunities.

THE HANDY TOOL

The basic toolkit is named after Olduvai Gorge in Tanzania, where Louis and Mary Leakey first found simple stone artifacts. For about a million years, Oldowan tools served the purposes of the earliest *Homo* populations. About 1.7 million years ago, a major technological change began as hominins developed the strength and skill to strike large, oval-shaped flakes, typically 20 to 40 centimeters (8 to 16 inches) across. By 1.6 million years ago, the edges of these large, portable flakes were shaped by knapping around the circumference, thus making a tool that was usually pointed at one end and rounded at the other. These tools are called handaxes, and the tradition of toolmaking that began with them is known as the Acheulean, after the site of St. Acheul in France where handaxes were first discovered.

Characterized by pointed handaxes, bevel-edged cleavers, and other large cutting tools, the Acheulean lasted until about 400,000 to 250,000 years ago, possibly even a little later in a few regions of Africa and Europe. The persistence of handaxe toolmaking for 1.2 million years, coupled with the ongoing utility of the basic core-flake-hammerstone toolkit, means that the Acheulean encompassed an almost unimaginably long era. For the first million years—from about 1.6 million to 600,000 years ago—there is little evidence of any progressive change in Acheulean large cutting tools. There is variation and some improvement, depending on the quality of rocks chosen by Acheulean toolmakers; but handaxes that are 600,000 years old can look as crude, or as well made, as those a million years older. The impulse to innovate had apparently yet to enter the minds of *Homo erectus* and *Homo heidelbergensis,* who were responsible for making handaxes during that period.

The toolmakers could turn almost any type of rock into a handaxe. At what is now the Smithsonian research site of Olorgesailie in Kenya, early handaxe makers

Using a scanning electron microscope (SEM), researchers can distinguish cut marks on ancient bone (bottom) by comparing them with cut marks made using stone tools on modern bone (top).

had access to at least 14 different types of volcanic rocks, all of different quality in terms of breakage and durability. Our studies show that toolmakers really knew their rocks. Mechanical engineering tests indicate that the Olorgesailie hominins most often picked the two finest types of rock—those that were the most brittle, and thus most easily flaked; and the toughest, possessing the most durable edges. These two rock types were transported farther than any of the others. The handaxe makers had a fine intuitive knowledge of rock mechanics, an ability to identify rocks of the best quality, and an astute spatial map of the landscapes where these rocks were found. Their persistence in making large cutting tools for a million years, however, makes it seem like human evolution went into a long stall.

What was so great about handaxes that they remained central to early human technology for such a long time? To answer this question, it is important to see things not from our present-day perspective, when technologies change so fast and depend on cultural know-how communicated by language. For many decades, archeologists thought that language must have been in place in order to maintain the handaxe tradition for tens of thousands of generations. But one can't help think that if language had been vitally important in teaching how handaxes should be made, the toolmakers must have said the same thing over and over again for a very long time. As we will explore later, one of the hallmarks of language is its mutable quality—the way it allows new ideas to be expressed and innovations to occur. The hallmark of the handaxe makers was a sense of stability, at least with regard to things we know they made.

Part of the answer comes from understanding the improvement that handaxes offered over earlier Oldowan technology. The starting point for Oldowan toolmakers involved rounded cobbles or odd-shaped chunks of rock, which

The earliest known stone tools significantly revamped the ecology of human ancestors, leading them to exploit large mammals that were rich in protein and fat.

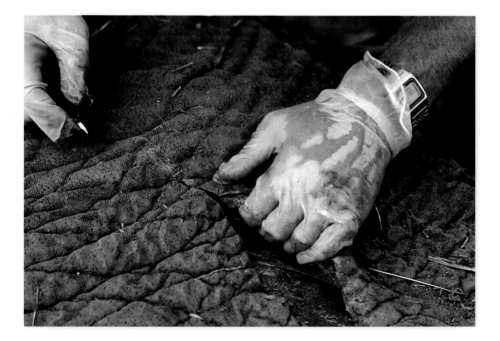

Researchers test a stone tool on elephant hide. Microscopic patterns of wear on the stone help establish the uses of ancient tools.

Thousands of handaxes litter the landscape at Olorgesailie in Kenya, where Smithsonian Institution paleoanthropologist Rick Potts conducts research on the lives and adaptations of early humans.

meant that a lot of wasteful chips and debris resulted as the acute edges of cores were created and freshened up and flakes were produced. By first knocking off a large flake and thinning its thickest part, the handaxe makers could walk away from the outcrop with a large rock that already had a very usable edge all the way around, and could be struck to produce hundreds of flakes with much less waste. Handaxes, therefore, were handy devices to carry around, enhancing the mobility of Acheulean foragers.

Another clue to the success of the handaxe makers comes from looking at the conditions in which they lived. Working at Olorgesailie, the Smithsonian research team has documented extensive changes in landscapes, food, and water resources that Acheulean populations experienced for hundreds of thousands of years. This dynamic environment suggests that Acheulean handaxes were not a symptom of stagnation but rather were part of a flexible repertoire of behavior. Handaxes were like a Swiss Army knife of the Stone Age—portable, quick to use, and a convenient source of flakes, with a lengthy, serviceable edge ready to be knapped at a moment's notice. It was useful for tasks ranging from delicate cutting to hacking through tree branches; modern-day experiments show the wide variety of tasks that handaxes could accomplish. Anvil stones were also part of the Olorgesailie toolkit, and were used to pound tubers and crush pods of many different food plants. The Acheulean toolkit could be used in the different situations and changing landscapes the toolmakers encountered.

FIRE

Before the onset of handaxe technology, early hominins were able to disperse from Africa to Asia armed only with Oldowan technology. Even the simplest stone tools helped early humans adjust to unfamiliar environments. By around 1.2 million years ago, a second expansion of hominins from Africa carried handaxe technology to western and southern Asia. Around 800,000 years ago, large cutting tools made an appearance in eastern Asia. In the Bose Basin of southern China, large-cutting-tool technology helped early humans colonize a landscape that was recovering from a nearby meteorite impact. By 500,000 years ago, Acheulean handaxes also made their way to the cold habitats of northern Europe, including Britain.

As human ancestors reached new places, change was slow, but the occasional innovation opened up new possibilities. Most important was the control of fire. Small patches of burned earth or fragments of burned bone occur in a few sites as much as 1.8 million years old, but there is little consensus as to whether humans could control fire at that time. The oldest definitive evidence of hearths is around 790,000 years old and was found at the site of Gesher Benot Ya'aqov in Israel. Concentrations of burned seeds, wood, and tiny pieces of flint from toolmaking activities near the fire signal a significant development in the lives of early humans. The control of fire at a hearth gave our ancestors the ability to cook food, find warmth and safety from predators, and share food and information with others in the group. Cooking made many foods more digestible and increased the amount of energy that certain foods could provide.

The expansion of the toolmakers into new regions placed demands on the body that favored anatomical adaptations to warm and cold climates. The evolution of hominin bodies continued to be important, although tools and fire began to provide new cultural ways of adjusting to the survival challenges ahead.

Handaxes were like a Swiss Army knife of the Stone Age—portable, quick to use, a convenient source of flakes with a lengthy, serviceable edge ready to be knapped at a moment's notice.

FAQ:
How do we know early humans started to eat more meat?

Archeological sites often preserve evidence of butchery of large animals, which was possible by using sharp stone tools. Such sites typically preserve numerous tools and signs of butchery, including bones with cut marks and some that were broken open to get to nutritious marrow. This type of evidence documents the shift from a plant-based diet to a more meat-based one.

By studying butchery sites, archeologists can see shifts in favored types of prey and in the tools used to process carcasses. Later species of early humans used weapons to hunt animals and fire for cooking. Both innovations made even more types of food available for consumption, including meat from larger and more dangerous prey.

CHAPTER 7

HUMAN PROPORTIONS

HUMANS LIVING TODAY ARE REMARKABLY VARIED IN BODY SHAPE. WE CAN BE SMALL bodied, like the Batek of Malaysia, or tall and slender, like the Maasai of Kenya. We can also be wide in the hip, which suits us well to life at high altitudes or latitudes. These shapes are nothing new. The human lineage evolved body proportions like ours hundreds of thousands of years ago, and there were many more variations before that. The human body has changed remarkably over time. What were the steps along the way?

Humans come in all sizes and shapes, and many of us feel our bodies are far from ideal. Nevertheless, despite variations in muscle, fat, and skin, the proportions of the human skeleton are relatively constant across our species. However, *Homo sapiens* possesses only the most recent model of many that evolved over the past six million years.

Most of those earlier bodies had proportions different from ours.

Our human ancestors had one of four general body shapes. The first was that of the early hominins, best known from the 3.2-million-year-old australopith known as Lucy. This body type combined ape and human proportions. The second was that of early *Homo,* best represented by the 1.5-million-year-old "Turkana Boy" skeleton, which shows early signs of modern human proportions. Third was the figure presented by *Homo heidelbergensis,* a species that evolved by at least 700,000 years ago, and also the 200,000-year-old Neanderthals. The latter two species were proportioned for tough lives and cold environments. Finally, there is the modern human form, which is much less robust than some of our ancestors' bodies.

LUCY AND HER KIN

How could one tell an early hominin from a chimpanzee if they were standing side by side? Lucy would have been quite similar to a chimpanzee in overall body size and shape. As we saw in Chapters 4 and 5, however, signs of her humanity—a body built for walking bipedally and shortened canine teeth—are also apparent.

Lucy was a stocky adult female just over a meter (3 feet, 6 inches) tall and weighed a solid 29 kilograms (64 pounds). Her head would barely have reached

Opposite: Long legs and relatively short arms of a gymnast reflect traits that evolved in the human lineage nearly two million years ago.

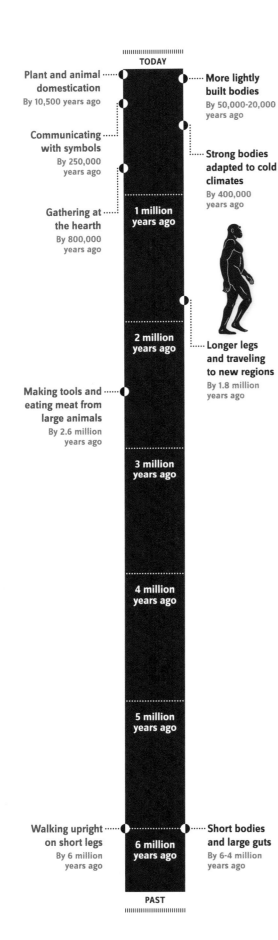

TODAY

Plant and animal
domestication
By 10,500 years ago

**More lightly
built bodies**
By 50,000-20,000
years ago

Communicating
with symbols
By 250,000
years ago

**Strong bodies
adapted to cold
climates**
By 400,000
years ago

Gathering at
the hearth
By 800,000
years ago

**1 million
years ago**

**2 million
years ago**

Longer legs
and traveling
to new regions
By 1.8 million
years ago

Making tools and
eating meat from
large animals
By 2.6 million
years ago

**3 million
years ago**

**4 million
years ago**

**5 million
years ago**

Walking upright
on short legs
By 6 million
years ago

**6 million
years ago**

**Short bodies
and large guts**
By 6-4 million
years ago

PAST

a doorknob. Her legs were more or less the same length as those of a chimpanzee of the same height, whereas her arms were shorter. Still, she had longer, stronger forearms than later humans would; they were reminiscent of an ape's powerful upper body, which is useful in climbing trees. The combination of long arms and short legs caused Lucy's hands to extend farther down her thighs than ours do. Were it not for the fact that her hands were like ours, and not long-fingered like a chimpanzee's, her fingertips might very well have touched her knees.

Two other features shared by Lucy's kind and chimpanzees are the shape of the rib cage and the setting of the shoulders. Apes' shoulders are set high relative to the head and close together, making it seem as if apes have no neck. Lucy's shoulders were not set as high, but they were narrow. This, combined with a rib cage that was wide at the bottom and narrow at the top, would have given her a broad, apelike torso and provided ample room for a large gut with a long intestinal tract. The implication is that the diet of australopiths, like apes, was largely plant-based. A long intestinal tract means that plant foods, which require more time to digest effectively than meat or insects do, could pass slowly through the gut.

Despite the similarities to chimpanzees, the details of her posture make it clear that Lucy was no ordinary ape. From her feet to the base of her skull, her skeleton defines a humanlike pose. A chimpanzee's foot has a big toe that diverges from its other toes and functions like a thumb, but Lucy's toes were aligned and faced forward like ours do. Where a chimpanzee's short legs are bowed, Lucy's knee could straighten fully like ours and was built to support the full weight of her upper body.

Lucy's pelvis is wide and bowl-like and unquestionably human in form when compared with the elongated, bladelike pelvis of a chimp. The key to the differing postures of Lucy and a chimpanzee is the spine. A chimpanzee's spine forms a continuous arc from its pelvis to the back of its skull. This suits an ape's four-legged, knuckle-walking posture. An S-shaped spinal column, by contrast, connects from Lucy's pelvis to the base of her chimp-size skull. This dramatically altered spine shows that australopiths habitually walked on two legs with heads held high—something apes do not do.

Despite the similarities to chimpanzees, the details of her posture make it clear that Lucy was no ordinary ape.

PROPORTIONS LIKE OURS

When did body proportions begin to shift away from the persistent mixture of apelike and human proportions that characterized the early hominins? An early clue comes from the site of Bouri in the Middle Awash area of Ethiopia. In 2.5-million-year-old layers, researchers found evidence of a hominin species whose femur was proportionally longer than Lucy's. While its legs had evolved a greater length, its arm bones retained apelike proportions. In the same layers researchers found the butchered bones of an extinct wildebeest and horse, and 278 meters (304 yards) away, the skull of an australopith, named *Australopithecus garhi*. Since the tool-marked bones and skull were found some distance away from the limb bones, one cannot be sure exactly how these three tantalizing pieces of evidence are related. Whether it was a late australopith who first began to use tools, eat meat, and have slightly lengthened legs is a question that only further discoveries may answer.

Between 2.5 million and 1.5 million years ago, a lot of variation characterized the hominins as the genus *Homo* was evolving from australopith ancestors. Skulls suggest that several species of early *Homo* lived during this time. By about two million years ago, one of these—probably *Homo erectus*—had evolved a tall, long-legged body, evident from sturdy thighbones and a powerfully built hip bone found in the fossil beds of Lake Turkana in Kenya.

The fullest early expression of proportions like ours appears in the *Homo erectus* skeleton known as the Turkana Boy, which comes from layers at Lake Turkana around 1.5 million years old. Although he was around eight years old when he died, the Turkana Boy was already 1.57 meters (5 feet, 2 inches) tall. A female

Opposite: *Muscular arms and a large, apelike upper torso probably contributed to the stocky appearance of Lucy, a 3.2-million-year-old* **Australopithecus afarensis** *who stood about a meter (3 feet, 6 inches) tall, as reconstructed by artist John Gurche.*

Australopithecus afarensis
AL 288-1, Lucy
About 3.2 million years old

Homo erectus
KNM-WT 15000, Turkana Boy
About 1.5 million years old

Homo neanderthalensis
La Ferrassie 1 and Kebara 1
About 70,000 to 60,000 years old

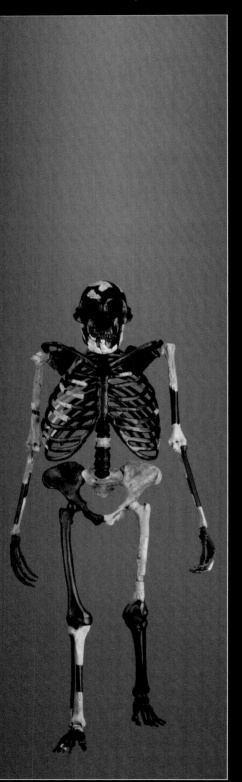

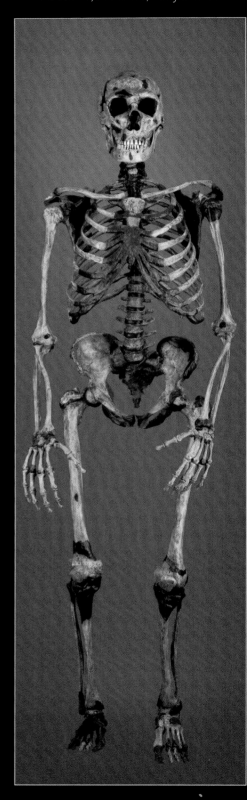

*Very short stature, long arms,
and short legs characterized*
Australopithecus afarensis.

The skeleton of this young **Homo erectus**
*shows the long and lean proportions
characteristic of humans who are adapted
to hot environments.*

*Though descended from a taller ancestor,
Neanderthals evolved a short, stocky body
to cope with cold Eurasian winters.
This body type helps retain heat.*

known from a more fragmentary skeleton of the same species is estimated to have been as tall as 1.8 meters (5 feet, 11 inches). Early African *Homo erectus* had thus evolved a tall, narrow physique in contrast to Lucy's stout one. The legs of the Turkana Boy were elongated to modern human proportions, as were his upper and lower arms. Although the length of his legs indicates that long-distance walking and running were a key part of *Homo erectus*'s suite of behaviors, this particular youth's vertebrae were slightly misaligned, pointing to a spinal disease that may have caused his early death.

The Turkana Boy had shed the wide, funnel-shaped rib cage of the australopiths and evolved a narrow, barrel-shaped thorax. His pelvis was also proportionally narrower than the australopiths'. This implies a shift in *Homo erectus* toward a short gut, which indicates that *Homo erectus* had adapted to a different diet.

Since the time of the australopiths, meat had become an increasingly important part of the human diet because it provided more "bang for the buck" nutritionally than plants did. But meat is not particularly easy to digest. Eating more meat required evolutionary changes. By the time of *Homo erectus,* if not earlier, our large intestines had become shorter and our small intestines longer, reducing the bulk of our gut as well as the time it took to digest food. Our guts became more like those of big cats, which rapidly gather nutrients from meat and then get rid of waste. Instead of taking days to digest plants, as gorillas do, our digestion is measured in hours.

The shift toward shorter guts allowed human bodies to divert precious energy from the intestines to any of the other metabolically expensive organs in the body. The most expensive of them all is the brain. The brain of early *Homo erectus* was roughly twice the size of an australopith's. The dietary shift toward meat and the reduction in gut size are likely to have proved crucial in fueling the evolution of an enlarged brain.

The long limbs and tall, narrow body of early *Homo erectus* in Africa were also an important adaptation to warm weather. Several groups of modern humans living in eastern Africa today, such as the Maasai of Kenya and the Dinka of southern Sudan, have similar physiques. In a warm climate it would have been an advantage to shed as much heat as possible by sweating. A narrow body with long arms and legs helps accomplish this by increasing the exposure of skin to air. The more surface area available for perspiration, the more cooling can be achieved as the sweat evaporates. *Homo erectus* was perhaps also among the earliest hominins with reduced body hair and an increased density of sweat glands as part of an evolutionary response to a warm climate. The slender body and long limbs also suggest that this species may have been the first to spend much of the daytime walking and foraging for food in the hot equatorial sun.

ADAPTING TO NEW ENVIRONMENTS

The oldest definite evidence of hominins outside of Africa is from Dmanisi in the Republic of Georgia, a site with fossils of *Homo* and stone tools dated 1.78

Reconstructing human ancestors

Visually representing extinct early humans is an art that requires exceptional understanding of human and ape anatomy. Artists who specialize in this field dissect human and ape cadavers to study and measure muscle attachments, fat, and the amount of tissue over bone.

When some bones from a fossil specimen are missing or incomplete, which often happens, the artist usually works with a scientist to fill in the blanks. Using similar bones from other specimens of the same species as a guide, they scale them up or down as necessary.

Still, some aspects of a reconstruction fall into the realm of informed speculation. We may never know, for example, when humans evolved the whites of the eyes, or what the skin color or hair texture of our ancestors was like.

The large digestive tract of **Australopithecus afarensis** *(left), indicated by the region between the ribs and pelvis, was helpful in digesting a plant-based diet. Meat could be digested quickly in the smaller digestive tract of* **Homo erectus** *(center).* **Homo heidelbergensis** *(right) depended on raw and cooked food that was efficiently processed in a short digestive tract.*

million to 1.75 million years old. To the east, stone tools and fossils from China and Indonesia indicate that hominins arrived there a short time later, by about 1.7 million years ago.

This geographic expansion implies that early *Homo* could adjust to new climates and ecosystems. The earliest groups to explore other continents did so with the simplest of toolkits—hammerstone, core, and sharp flakes that typified the oldest known stone technology. This suggests that even the most basic adaptations of the genus *Homo* made this early wanderer capable of expanding to new habitats.

Although not all scientists agree on which species this was, the fossils from Dmanisi have an overall appearance like *Homo erectus.* Two upper incisor teeth from the site of Yuanmou in southern China, dated 1.7 million years old, are a close match to those of the Turkana Boy. The first known hominins to reach

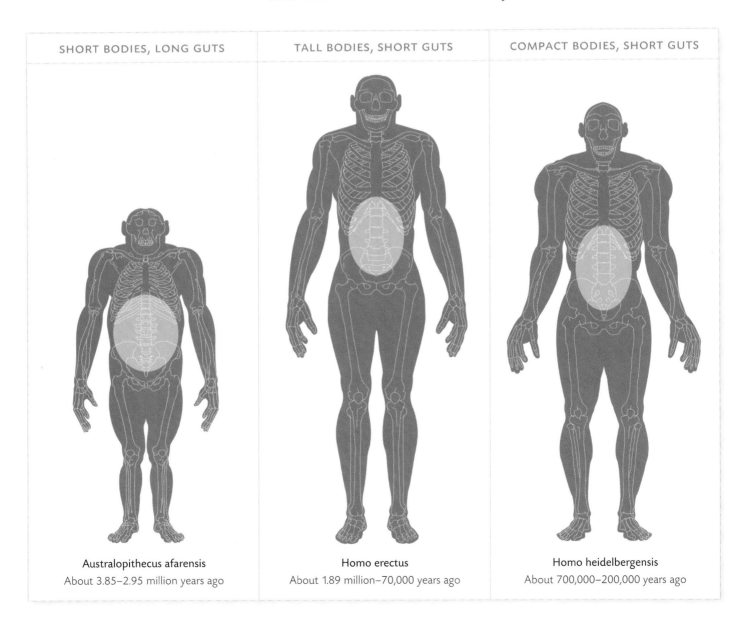

SHORT BODIES, LONG GUTS	TALL BODIES, SHORT GUTS	COMPACT BODIES, SHORT GUTS

Australopithecus afarensis
About 3.85–2.95 million years ago

Homo erectus
About 1.89 million–70,000 years ago

Homo heidelbergensis
About 700,000–200,000 years ago

Java, Indonesia, at least 1.66 million years ago, have long been assumed to be *Homo erectus,* like the later populations of the island; yet the fragmented condition of this oldest sample leaves the matter open to interpretation. A more remote line of evidence comes from late in the story: The species *Homo floresiensis,* nicknamed the "hobbit," had body proportions and other characteristics that look like African hominins prior to *Homo erectus.* Could it be that a species moved out of Africa and across Asia even before the tall, lean, and powerful *erectus?*

It's generally thought that the oldest colonizer must have been, like *Homo erectus,* capable of endurance walking and was, possibly, a meat eater who required large areas to successfully scavenge and hunt for animal protein and fat. During the crucial interval between 1.9 million and 1.7 million years ago, East Africa experienced marked fluctuations in moisture and lake levels. Groups of hominins may have expanded along lush valleys and up the Nile during moist times but later have been unable to retrace their path when drought closed in behind them. Perhaps there was no direction to move but north.

By 1.8 million years ago, colonists had reached northwest Africa. At the site of Aïn Hanech in Algeria, they encountered a broad, grassy plain carved by a network of streams with marshes, shrubland, and forests nearby. To the northeast, at the foot of the Caucasus Mountains, they occupied wet river valleys and forests mixed with open woodland, dry meadows, and rocky outcrops. Across north-central India and into southern China, grassy terrain supporting herds of animals was common. Woodlands and forests were also within sight. By 1.7 million to 1.6 million years ago, hominins had expanded all the way from the cool, dry grasslands and woods of the Nihewan Basin in China, at 40° N latitude, to the warm, humid plains and wooded wetlands of Java at 7° S. As they ventured along the narrow peninsulas and land bridges of Southeast Asia, these colonists also encountered lagoons, seaside deltas, and beaches. An ability to adapt to varying survival conditions was already ingrained in this ancestor of ours.

Tracking his quarry in frigid conditions, an Inuit hunter conserves heat through a short, stocky body shared with other people who live in cold climates. Humans in the tropics, by contrast, have evolved tall, thin physiques with more skin surface.

INVADING COLD ENVIRONMENTS

Between about 1.3 million and 1 million years ago, hominins reached western Europe and moved as far north as the British Isles, leaving behind stone artifacts but very few of their own bones. Fragmented fossils from the Gran Dolina cave

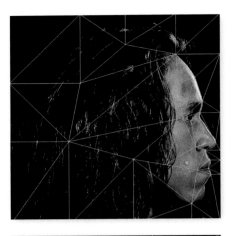

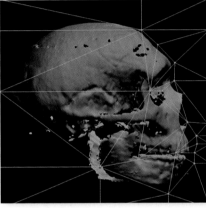

Differences in appearance between Neanderthals and modern humans become very apparent when researchers using image manipulation software take key anatomical points on a Neanderthal skull (center) to transform a modern human model (top) into a Neanderthal (bottom).

sites in the Sierra de Atapuerca of Spain are 1 million to 800,000 years old, but the diagnostic bones that could identify the colonists as *Homo erectus,* earliest *Homo heidelbergensis,* or *Homo antecessor*—a different species altogether—have not yet been found. We do know, however, that these human ancestors experienced cold seasons every year, even though the largest advances of glaciers were yet to come.

An increase in body mass, the strength of the limb bones, and the breadth of the pelvis are registered in *Homo heidelbergensis* in Europe between 700,000 and 300,000 years ago. The larger, wider body of this species contrasts markedly with the narrower build of early African *Homo erectus* and indicates a body evolved to withstand cold temperatures. The impressively robust bones also suggest a degree of strength unmatched in earlier hominins or in ourselves.

As *heidelbergensis* became concentrated in Europe, *Homo erectus* continued on in East Asia. African hominins from about 600,000 to 250,000 years ago look similar, overall, to their European contemporaries. With time, distance, and adaptation to different environments, however, European and African populations eventually diverged. The European branch led to the Neanderthals; the African, to the immediate ancestors of our own species, *Homo sapiens.*

It is in the Neanderthal lineage, from around 200,000 to 28,000 years ago, that we see the clearest signs of a way of life focused on surviving in cold climates. Slightly shortened forearms and lower legs and a wide, robust torso served effectively to conserve heat. Neanderthals also had enormous noses with expanded nasal membranes that played, literally, a life-or-death role in warming and moistening cold, dry air as it was inhaled—and then recapturing the moisture inside the nose before the air was exhaled. This spared the lung damage that would have occurred with sustained breathing of frigid air.

Like *heidelbergensis,* the Neanderthals lived through several intense ice age cycles, when glaciers expanded across the north and downward from the high mountain chains of Europe. Yet these cycles were exactly that—wide swings between cold and warm conditions. There were times so balmy that hippopotamuses migrated from the tropics into present-day England. While Neanderthals experienced warm periods, their bodies were built for the trials of the cool seasons and the rarer, harsher ice ages that occurred during their tenure in Europe, the Near East, and Central Asia.

All told, the Neanderthals appear to have been a tough, muscular species. One study comparing their injury patterns to modern humans' found that the Neanderthals' injuries most closely resembled those of modern-day rodeo riders, who sustain multiple upper body fractures and head injuries. Animal bones found near Neanderthal hearths and burial sites suggest that their meat-dependent diet required them to hunt large prey such as aurochs, rhinoceroses, and bears—animals that could injure them, especially at close quarters. Muscular bodies with tough bones would have been a distinct advantage.

A MORE DELICATE SPECIES

Neanderthals retained much of the brawn and robustness of our common ancestor, *Homo heidelbergensis*. In our own species, however, a distinct trend toward smaller and weaker skeletons occurred. Although human populations are often considerably taller than Neanderthals, we evolved thinner bones and less massive bodies. Compared with *Homo heidelbergensis* and Neanderthals, we are rather delicate.

This light build is apparent in our heads as well. Our faces are small and do not have the jutting jaws of *Homo heidelbergensis* or Neanderthals. Our brow ridges, if they are apparent at all, are greatly diminished compared with these other hominins'. While Neanderthal jaws are deep and solidly built, ours are shallow and lightly built, and we are the first hominin to possess a peculiar projection called a chin. It came about when the part of the jaw that holds the teeth and tooth roots evolved a smaller size. When that happened, the buildup of bone that strengthens the midline of the jaw could no longer project inward, as in all earlier hominins; it could only grow forward—thus, the chin.

Our globe-shaped heads and small, flat (and in females, hairless) faces may even have a parallel in domesticated animals, whose breeding has produced more delicate bones and infantlike faces in comparison with the adults of their wild ancestors. In the case of dogs, for example, humans have effectively bred the wolf out of most breeds. Why the strength and ruggedness of ancient ancestors has been bred out of humans is unclear. One possible explanation is that a more delicate appearance may have stimulated a caring or nurturing response.

The Neanderthals were physically adapted to survive the cold, but modern humans conquered it and many other environments around the globe through technology, hunting strategies, innovation, and social and trading networks. As modern humans experienced further changes in lifestyle—such as the shift to planting crops and a daily regimen of less demanding activity—over time, the evolutionary benefits of brawnier bodies never reappeared.

The thickness of the bone of a Neanderthal femur (left), compared with that of **Homo** sapiens *(right), suggests the former's rugged lifestyle called for strong bones.*

FAQ:

What's so controversial about the "hobbit"?

Recent discoveries on the Indonesian island of Flores have shown evidence of the species *Homo floresiensis* (nicknamed "hobbits"), which became extinct by just 17,000 years ago. This species displays body proportions more similar to Lucy's than to the Turkana Boy's (and modern humans') and raises interesting questions about the evolution of body proportions in hominins.

Is it possible that a body form similar to much earlier hominins' reevolved as a consequence of "island dwarfing," by which animal species shrink over time when isolated on islands? Or did a lineage of hominins maintain ancestral body proportions that were otherwise thought to have disappeared 1.5 million years ago? More discoveries may help resolve this ongoing debate.

CHAPTER 8

EVOLUTION OF OUR BRAIN

THE HUMAN BRAIN IS THE SOURCE OF INTELLECT AND INSANITY, CREATIVITY AND CRUELTY. It's the place where the powers of belief, reasoning, and emotion meet in a compact universe of a hundred billion firing neurons. Crucial to our ancient survival, the human brain tripled in size as we evolved. Always hungry, it is fueled by whatever energy we consume. When our enlarged heads enter the world, the trials of birth make it a wonder that humans ever survived. Fortunately, no one's brain works alone. With nurturing and encouragement, we thrive and can be calmed.

Far from the African cradle of humanity, there lived an early human destined to make scientists rethink certain assumptions about human origins. A fossil skeleton of this new hominin, nicknamed "hobbit," was excavated in 2003 at Liang Bua, a cave on the island of Flores, 625 kilometers (388 miles) or so east of Java in Indonesia. Hominin fossils from the cave were soon established as a new species, *Homo floresiensis.* Like Lucy, the "hobbit" skeleton was an adult female. She was short—at only 1 meter (3 feet, 3 inches) tall, even a little shorter than Lucy. And she had a Lucy-size brain of about 420 cubic centimeters. But *Homo floresiensis* lived much more recently, from about 95,000 to 17,000 years ago.

The scientific world was puzzled by the diminutive size of this species, and especially by its brain. Researchers have long observed that humans evolved larger brains over time. So why was *Homo floresiensis*'s so tiny? Did its body and brain become reduced for some reason, or did it arrive on the island small and stay that way? The "hobbit" was also associated with behaviors—toolmaking, controlling fire, hunting game, and fending off large predatory lizards—far beyond what a small-brained hominin was thought capable of. How could *Homo floresiensis* do these things? How much brain is needed to power these activities? The conundrum of the "hobbit" focuses new attention on what the human brain is all about, and how it evolved.

The brain of a chimpanzee is roughly 400 cubic centimeters, whereas the human brain is about 1,350 cubic centimeters, more than a threefold difference. The outer portion of the brain—the cerebral cortex—is mainly responsible for the huge difference in size between human and other primate brains. The deep

Opposite: *Electrodes attached to the head of a Tibetan teacher help neuroscientists measure the activity of the human brain.*

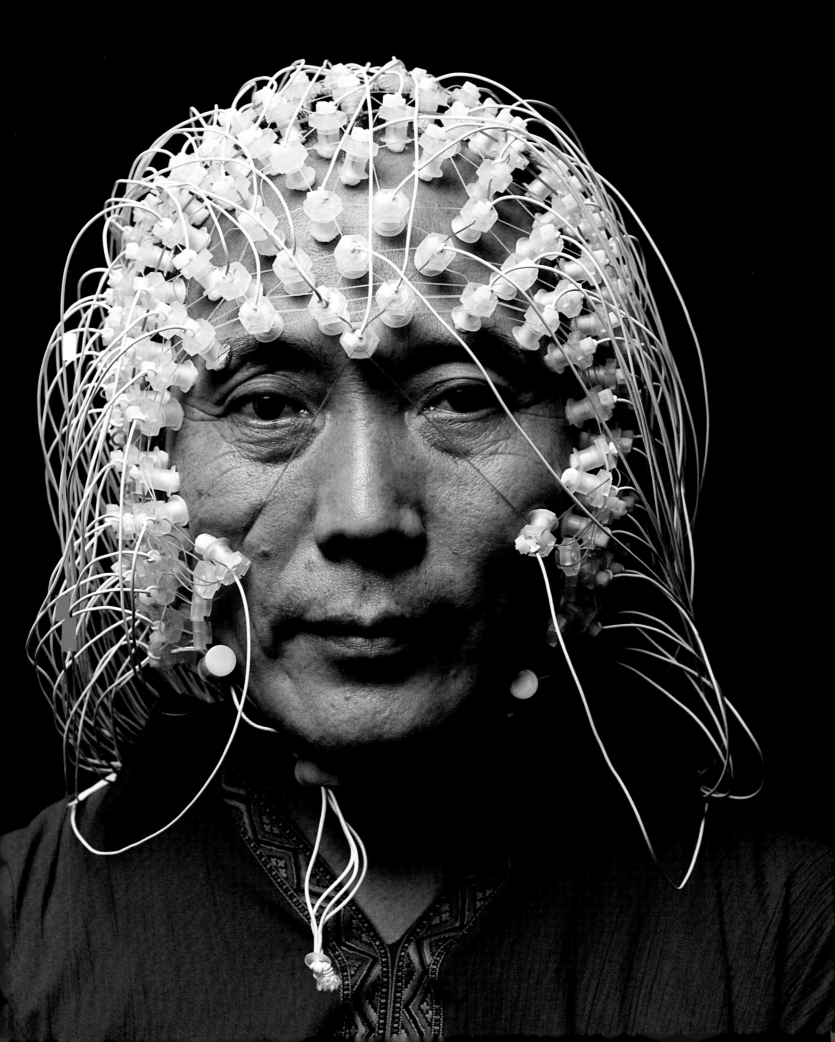

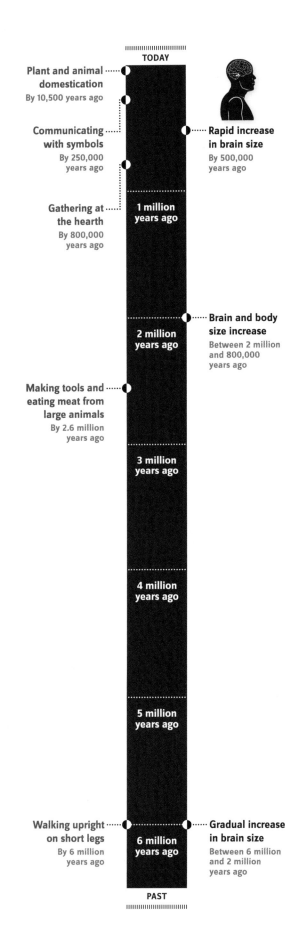

TODAY

Plant and animal domestication
By 10,500 years ago

Communicating with symbols
By 250,000 years ago

Gathering at the hearth
By 800,000 years ago

1 million years ago

Making tools and eating meat from large animals
By 2.6 million years ago

2 million years ago

3 million years ago

4 million years ago

5 million years ago

Walking upright on short legs
By 6 million years ago

6 million years ago

PAST

Rapid increase in brain size
By 500,000 years ago

Brain and body size increase
Between 2 million and 800,000 years ago

Gradual increase in brain size
Between 6 million and 2 million years ago

crevices visible on the external surface divide the cerebral cortex into its main regions, or lobes, which are divided further based on the microscopic patterns of nerve cells and the connections among them. Areas where cell bodies are packed together appear darker and are called gray matter, while the projections that connect nerve cells constitute what is called white matter—which forms the communication networks of the brain. Compared with those of other primates, the human brain is also unusual in its internal wiring. Our enlarged neocortex, evolutionarily the most recent part of the cerebrum, exhibits many more connections among nerve cells than a chimpanzee's does. This is evident in an expansion of white matter in vital areas of the cortex. The parts of our brains concerned with higher mental processing are better connected and faster at processing information. As much as 75 percent of all connections in the human brain are within the cortex, enabling it to process new information and deliver output in split seconds, integrate decades' worth of stored information, and create images, interpretations, and abstractions. Wiring is thus very important in human brain evolution—a point that might help explain how *Homo floresiensis* could live in ways that, at first, seem more sophisticated than its brain size could support.

MEANING OF THE BRAIN

Nineteenth-century neurologists discovered that particular brain regions control particular behaviors, including the ability to speak and comprehend language. Ever since, scientists have sought to study the brain's structure and its workings. An assortment of technologies have opened up new possibilities for uncovering the links among brain anatomy, brain activity, and the localization of those mental activities unique to humans. Increases and decreases in brain activity can be measured by PET (positron-emission tomography) scanning, which offers a three-dimensional record of brain metabolism, and by functional MRI (magnetic resonance imaging), which shows blood flow in the brain during particular mental activities. As we learn more about it, brain functioning may someday replace the idea of intelligence as the gauge of human aptitudes.

As self-conscious creatures, we may find it difficult to imagine that vital parts of our identity can be linked to discrete bits of gray and white matter. Yet time and again it has been demonstrated to be the case. The frontal lobe of the cerebral cortex offers a good example for two reasons. This area expanded a lot during human evolution, so it's a good place to look for the basis of uniquely human qualities. The frontal lobe is also the most commonly injured brain area due to its location at the forward part of the skull. Injuries and post-injury behavior indicate that the frontal lobe is associated with awareness of surroundings and the ability to plan. It plays a significant role in emotional responses, memory, and motor activities, including the ability to control small, complex movements of the hands, fingers, and facial muscles. One well-defined area is associated with expressive language. Due to this variety of functions, a frontal lobe injury can cause a myriad of problems, such as an inability to finish making a cup of coffee or difficulty finding the way home. People with frontal lobe damage, often caused by strokes, can show dramatic changes in social behavior. These include mood shifts, loss of a sense of humor, loss of empathy or sympathy, loss of sexual interest or the emergence of socially inappropriate sexual behavior, and the inability to detect deception.

Much of what we consider the core of our individuality—our memories, our ability to interact with others, even our ability to get a joke—depends on the functioning of our brain. Indeed the whole universe of human cultural phenomena—our belief and value systems, our complex social lives, the myriad complex calculations in the economic and scientific realms, the flourishing of creativity in the realm of art and imagination, all astonishing and uniquely human characteristics—would not exist were it not for the brain.

BRAIN SIZE

Despite the importance of the brain's internal connections and the functioning of its particular parts, the study of human brain evolution has mainly been informed by the interior of fossil braincases. If you graph the volume, or cranial

Hobbitlike proportions of **Homo floresiensis** *surprised researchers by showing that the evolution of brain size in our lineage was not always from smaller to larger.*

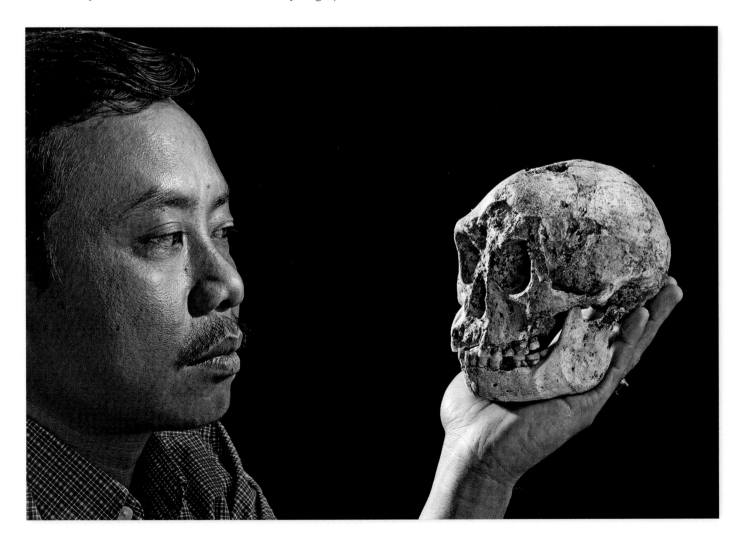

Climate Variability and Brain Size

Climate fluctuations (top), beginning around 800,000 years ago, corresponded with a rapid increase in hominin brain size (bottom). Larger, more complex brains enabled early humans of this time period to interact with other humans and with their surroundings in new and different ways.

capacity, of all available fossil braincases, the general impression is of a relentless progression toward larger brains during human evolution.

The cranial capacities of fossil humans dated between six million and three million years old mainly fall between 350 cubic centimeters and 500 cubic centimeters—mostly within the range expected for a great ape. To make sense of these numbers, the sizes of fruits like an orange, a grapefruit, and a small cantaloupe provide convenient visual markers. The earliest hominins had brains the size of oranges. The 350-cubic-centimeter cranial capacity of *Sahelanthropus* was exceeded by the brain capacity of the australopiths and *Paranthropus*. These later hominins usually had brains between 400 cubic centimeters and 500 cubic

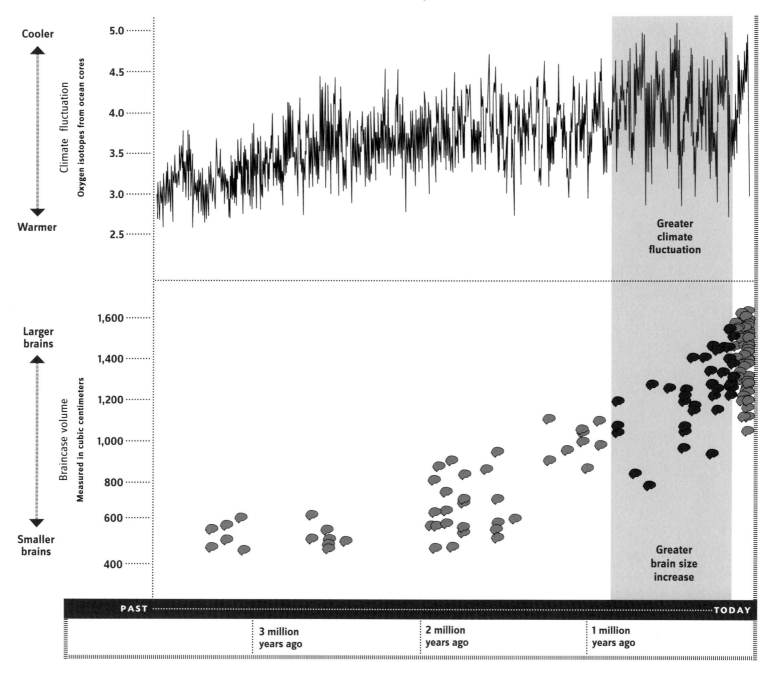

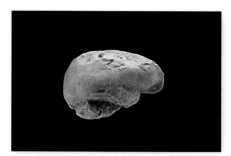

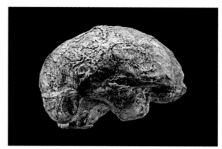

Australopithecus afarensis
3.1 million years old; 500 cubic cm

Homo rudolfensis
1.9 million years old; 775 cubic cm

Early Homo erectus
1.8 million years old; 850 cubic cm

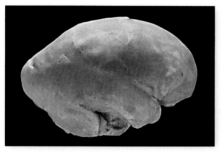

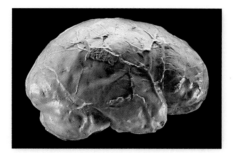

Homo heidelbergensis
350,000-150,000 years old; 1,200 cubic cm

Homo sapiens
26,000 years old; 1,322 cubic cm

Fossil brain endocasts provide researchers with hard evidence of how big our ancestors' brains were and some details of their surfaces.

centimeters; but a big *Australopithecus afarensis* male had one measuring 550 cubic centimeters. While the difference between 350 cubic centimeters and 500 cubic centimeters is considerable, the time span involved was more than three million years. Most researchers agree that brain size did not evolve rapidly in these early hominins.

Whether the brain functions of these early hominins differed much from those of our last common ancestor or even from living great apes is a difficult question to answer. Researchers predict animal cognitive ability by comparing the size of the brain with the size of a normal body, measured in weight. In this system, a large-bodied animal is expected to have a large brain, and a small animal is expected to have a small brain. If an animal has a smaller brain than might be expected for its size, then its intellectual capacity is predicted to be low. Conversely, if an animal has a larger brain than expected for its size, it would most likely be clever in ways we can test. Take, for example, the manatee—a slow-moving marine mammal with a large body and a small brain. A bottlenose dolphin, by contrast, has a large brain in proportion to its body. Until around two million years ago, when larger-brained hominins evolved, dolphins had the high-est brain-to-body size ratio in the animal kingdom. This ratio helps explain why bottlenose dolphins are capable of a myriad of complex behaviors, including tool use, while manatees do not appear to perform particularly complex mental tasks.

Applying this method to extinct humans is tricky because we cannot directly measure their body weight. So researchers have developed formulas based on living humans and apes to estimate body size and weight from fossil bones.

Although the accuracy of such estimates depends on the completeness of fossil skeletons, the calculations so far suggest that prior to 2.5 million years ago humans had a slightly larger brain than would be expected for an ape of their body size. These studies also show a slight increase in this ratio from the earliest hominins to the latest members of the genus *Paranthropus,* who lived until nearly one million years ago.

Increasing brain size is quite evident by 2 million to 1.5 million years ago. During this period, associated with the early evolution of the genus *Homo,* hominins with grapefruit-size brains evolved. The variation in brain size among these hominins was substantial, ranging from around 500 cubic centimeters, a size that overlaps with earlier hominins' brains, to more than 800 cubic centimeters in early *H. erectus.* The higher end of this range is equivalent to a grapefruit with a 12-centimeter (5-inch) diameter. The body weight of early *Homo* species increased during this period as well, perhaps as much as 60 percent. By 1.5 million years ago, average brain size had increased by at least 80 percent compared with that observed in three-million-year-old *Australopithecus afarensis.* These figures suggest that brain size slightly outpaced the increase in body size. By this time, *H. erectus* created new opportunities by using stone tools, seeking game, and eating a variety of foods. These strategies of meeting and adjusting to novel circumstances were probably powered by the extra increase in brain size.

Early *Homo erectus* reached a body size well within the range for modern humans. Thus, any significant increase in brain size after 1.5 million years ago would have added to neural and mental capabilities. The fastest rate of brain enlargement occurred between 800,000 and 200,000 years ago. This spurt produced the cantaloupe-size brains of later *Homo erectus* and *Homo heidelbergensis* and set the stage for the big-brained Neanderthals and modern humans. In this time interval, brain size increased by 50 to 70 percent, depending on which of the diverse species of *Homo* we look at. This enlargement occurred within only 600,000 years, and while hominins were robust, height and overall body mass leveled off during this interval.

In later *H. erectus,* a cranial capacity of 1,025 to 1,225 cubic centimeters was typical; and in *H. heidelbergensis,* 1,100 to 1,325 cubic centimeters was the norm. By 200,000 years ago, the lineage of the latter species had split into two—the Neanderthals and modern humans, both with cranial capacities that averaged around 1,450 cubic centimeters. However, *H. neanderthalensis* from Amud, Israel, had a capacity as high as 1,740 cubic centimeters, and *H. sapiens* from Cro-Magnon, France, one as high as 1,730 cubic centimeters. In our own species, average brain size has decreased slightly, along with lean body mass, compared with earlier fossil specimens.

OUR HUNGRY BRAIN

Size, of course, has its pluses and minuses, its costs and benefits, as do most developments in evolution. The human brain is no exception. At the top of the

Reading ancient brains

Complete fossil braincases are rarely found, but when they are, they provide a treasure trove of information. Brains hardly ever fossilize, but the interior of the braincase does bear an imprint of the brain's outer surface. A replica of the braincase interior, known as an endocast, can show the faint ridges and furrows on the brain's surface and indicate the relative proportions of different parts of the brain. Endocasts can also help estimate the brain's volume.

During fossilization, sediment may fill in the cavity where the brain used to be, forming a natural endocast. Scientists can also produce an artificial endocast by making a mold of the inside of a braincase. Medical imaging technology, such as computed tomography (CT) scanning, is now used often on fossil braincases to calculate cranial capacity, observe and measure internal structures, and produce digital casts of the brain.

list of costs is the fact that brains are very expensive to grow and maintain. In adults, the brain makes up about 2 percent of body weight, yet it uses 20 to 25 percent of the body's energy as measured by blood flow and the oxygen consumed during rest. Infants are born with brains only about 25 percent as large as an adult's, but about 60 percent of a baby's resting metabolism is devoted to its brain. Humans thus have an extremely hungry brain. It requires more energy than any other organ in the body.

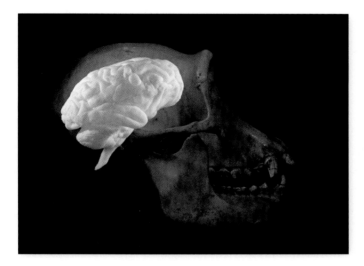 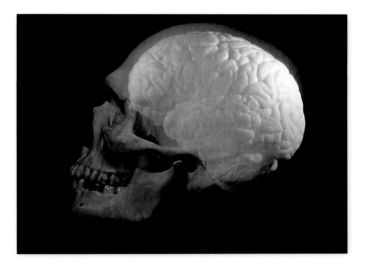

How does one feed it? Where does all the nourishment come from to support the rapid growth of a large brain from birth through childhood and to maintain the cerebrum's functions in adulthood? As we noted in Chapter 6, significant brain expansion placed a premium on finding foods with the highest nutritional value. Reliably acquiring a diet rich in animal fat and protein, ripe fruits, and large underground tubers—foods that promote and sustain brain growth—is a lot harder than grazing or eating leaves that can be found almost anywhere. A very large brain must be sustained by clever ways of capturing foods with the highest energy payoff.

A comparison of a chimpanzee brain (left) with a modern human brain (right) shows a marked difference in size. Yet the complexity and density of the connections among nerve cells is even more significant.

The risks of giving birth to a large-brained baby are another potential cost, one that can endanger both mother and child. After delivery, large-brained infants require an impressive investment of time and caring by others. Newborn babies, as captivating as they can be, are among the most incompetent mammals on Earth. It is vital to the whole enterprise of growing up and surviving that babies be able to prompt a deeply nurturing love. And since the brain takes more than a decade to mature, babies require an intensive, ongoing investment by parents and the immediate social group to improve their odds of surviving to adulthood.

If a modern-size brain presents so many problems from an adaptive standpoint, how could it have evolved to such a large size? A change in the diet, made possible when early humans began using stone tools to obtain meat, marrow, and other foods, was important in order for a hungry brain to evolve. But this explains how brains were allowed to expand, not necessarily what spurred the expansion in the first place.

Since both brain enlargement and a dependence on stone tools characterize *Homo,* it's conceivable that brain evolution and tools were somehow linked. Yet there isn't any clear association between an increase in brain size and the oldest stone toolmaking or with any particular change in stone technology. So tools alone don't make for a very compelling explanation.

Social interaction has been an important factor in primate brain evolution. In baboons and other Old World monkeys, larger group size correlates well with a

Tackling a puzzle, or exploring his world in any other way, a human child experiences very rapid brain growth, and rapidly multiplying neuronal networks, during the first few years of life.

bigger cerebral cortex. In most of the great apes, however, groups vary in size; individuals meet up and split apart on a regular basis rather than forage through the day in a stable troop. The fluid composition of the social group is especially evident in chimpanzees but also characterizes humans who hunt and gather for a living—or, in fact, most of us who work outside the home, send kids to school, and reunite later in the day after we part from our co-workers. This fission-fusion type of social grouping means that there's considerable advantage in retaining complex memories and in figuring out the mental states of others when you meet up after being apart. In brief, the flexibility of meeting and parting, along with the development of social alliances among groups and of broad networks of social communication, would have called upon a wide range of sophisticated brain functions.

Another factor in brain evolution is environmental complexity. As they dispersed to new places and had larger home ranges, *Homo erectus* and later humans encountered new habitats and challenges to their survival. Sophisticated mental imaging and the ability to integrate past memories with new inputs about place, time, and social group would all have been helpful in locating scarce foods that could make the difference between life and death in an unfamiliar environment.

Organisms can thrive only if they can respond successfully to seasonal fluctuations between wet and dry, warm and cold. Species can persist only if their constituent organisms can adjust to new resource fluctuations and other critical challenges that are outside of a single individual's memory. Vital functions of the cerebral cortex, such as accumulating experience, planning, and using imagination, were most likely beneficial whenever environments changed, thus spurring further brain evolution.

The fastest rise in overall brain size, which began around 800,000 years ago, corresponds with a time when global environmental changes became magnified. The instability of ancient landscapes can be seen in the strata of East African fossil sites, where large lakes and dry plains alternated over time in the places where early humans left tools and skeletal remains. The timing and duration of rainy and dry seasons were predictable and stable in some eras, but they were interrupted by unpredictable eras when rainy or dry seasons sometimes failed altogether.

In this dynamic setting, there were immense advantages to having a large brain with a well-developed cerebral cortex that functioned as a command center for integrating the senses, generating fine movements, reasoning about time and space, thinking, and verbal expression. The most anterior area, the prefrontal cortex, expanded more than other parts of the cerebrum. Its dense network of connections with other brain areas has led some researchers to suggest that its enlargement made humans capable of flexibility and novel responses. As the brain evolved, therefore, it became an organ of plasticity, enabling us to learn new things, make decisions, and create new associations and memories throughout our lives. In light of the ever changing challenges to early human survival, the brain evolved in ways that endowed our species with a unique adaptability.

> **As the brain evolved, it became an organ of plasticity, enabling us to learn new things, make decisions, and create new associations and memories throughout our lives.**

FAQ:
Brains: Does size matter?

Do bigger brains mean greater intelligence? Absolute size can be very misleading. Large animal species have big brains not because they are intelligent but because big brains are needed to operate large mammalian bodies.

While humans can be considered the most intelligent of species, we do not have anywhere near the biggest brains among animals living today. Whale and elephant brains are much larger than those of humans.

Brain size does matter, though, relative to body weight. By that measure, the larger the brain, the more intelligent the animal tends to be. For our body size, humans have the largest brains of any living organism.

OUR SOCIAL BRAIN

Understanding the workings of the brain remains one of science's greatest challenges and is part of an appreciation of what it means to be human. As we trace the evolution of the brain, we see that a variety of human ancestors—*Australopithecus afarensis, Homo erectus, Homo heidelbergensis*—contributed to the overall increase in size. Would we feel comfortable calling any of them human? Scientists used to argue over the existence of a threshold in brain evolution beyond which we could discern a species capable of thinking like us. Yet the most significant increments in brain size did not coincide with any one advance that defines us as human. Brain size evolved most rapidly long after our ancestors began making tools, for example, and well before any firm evidence of complex symbolic behavior appears in the archeological record. The ways in which human qualities became manifested over time, therefore, were not just a matter of brain size. Yet the enlarged brain continues to be important to many people's thinking about our identity as humans.

This question of human identity returns us to the mystery of the "hobbit." One implication of *Homo floresiensis*'s tiny head is that a progression in brain size over time was neither relentless nor inevitable. The downsides of having a huge brain are perhaps enough to convince us on the matter. When the options for feeding the brain become diminished, our proud emblem of mental superiority can become a liability. *H. floresiensis*, the "hobbit," faced exactly this situation on the small island of Flores in Indonesia. As a general rule, mammals that were once large tend to evolve considerably smaller brains and bodies after they reach isolated islands like Flores, Crete, and Madagascar. If the foods that could fuel the brain are unreliable, which is typical of a small island, evolving a hungry brain is a very risky enterprise. This doesn't answer the question of whether the tiny "hobbit" lineage became smaller over time or arrived on Flores that way. But it does imply that, with the hurdles set so high, there was nothing inevitable about brain enlargement over time during human evolution.

This point leads to a further insight when we consider our own humanness in an evolutionary perspective. Several defining elements of human behavior result from our concerted efforts to manage the consequences of an enlarged brain. The fact that we take so many years to grow up reflects the long time it takes for the brain to mature. Thus we put enormous energies into parenting, and we go to great lengths to obtain rich sources of food to help fuel the hungry brain. We also typically pool the efforts of gathering or shopping for food so that we can share and eat it with others. These economic and social features unique to human beings are all consequences of having evolved a large brain.

The brain, of course, rewards us. Many smaller-brained species rely on learning to get around and thrive. Chimpanzees and other great apes pass on learned traditions across generations, leading to behaviors unique to particular social groups—sometimes called great ape cultures. Humans, however, are by far the most reliant on cultural inheritance, which rapidly generates and communicates

> **Humans are by far the most reliant on cultural inheritance, which rapidly generates and communicates existing and new ideas and behaviors, all mediated by the brain.**

existing and new ideas and behaviors, all mediated by the brain. Our cultural abilities are often contrasted to our genetic inheritance, but the two are closely intertwined. Genetic inheritance acts on the storehouse of biological variation, of which only a small subset can be coded in the DNA inherited by any one individual. Since this kind of inheritance can be passed on only from parent to offspring, genetic responses to survival challenges usually require many generations to catch hold. During human evolution, though, our genetic inheritance has encoded the capacity to grow a large, adaptable, rapidly working brain that can instantly spawn new behavioral and mental possibilities. Our genetic inheritance includes the capacity for language, which is itself a complex code for generating almost limitless communications at short notice compared with the timetable of genes.

Opportunities for meeting the world in new and diverse ways grew exponentially when these two forms of inheritance, genetic and cultural, joined forces. The human brain became intensely social. The activities and thoughts of one brain became unavoidably entangled with many others. The brain is thus much more than a structure contained inside the skull. Our brains belong to the people around us as much as to ourselves.

A computer model of the brain's complex wiring illustrates the transmission of an electrical signal from one neuron to another. Billions of neurons form a network in each of our brains, which in turn link us in social networks that can include thousands of individuals.

A MEAL AT OLORGESAILIE

The Smithsonian Institution and the National Museums of Kenya have conducted research at Olorgesailie in Africa's Rift Valley since 1985. Since then, researchers have come to learn why the site, first surveyed and studied by Louis and Mary Leakey beginning in 1942, has one of the richest concentrations of Acheulean handaxes in Africa.

After spotting the end of an elephant's thighbone protruding from an eroded hillside, the Smithsonian research team dug to see if more of the fossil elephant could be found. Eventually most of the elephant's skeleton and more than 2,300 stone tools were found.

Several sharp grooves in a rib's surface were also found, signs that elephant meat had been sliced from the bone with sharp stone flakes. Stone tool butchery marks also occurred on other ribs, vertebrae, and even on the hyoid bone, where the elephant's tongue muscles had attached.

A clear picture began to emerge from dozens of sites nearby, in a layer about 990,000 years old. The elephant had provided a feast for early humans during an arid and difficult time. By making tools and cooperating with one another, humans were able to obtain a food source more than 150 times their own size. But who was the toolmaker? Along the route to the highlands, where the team had already identified stone quarries that the early humans visited to obtain material for their tools, a quick survey yielded the skull fragments of *Homo erectus*.

Stone technology and social cooperation were essential to the survival of early humans nearly a million years ago.

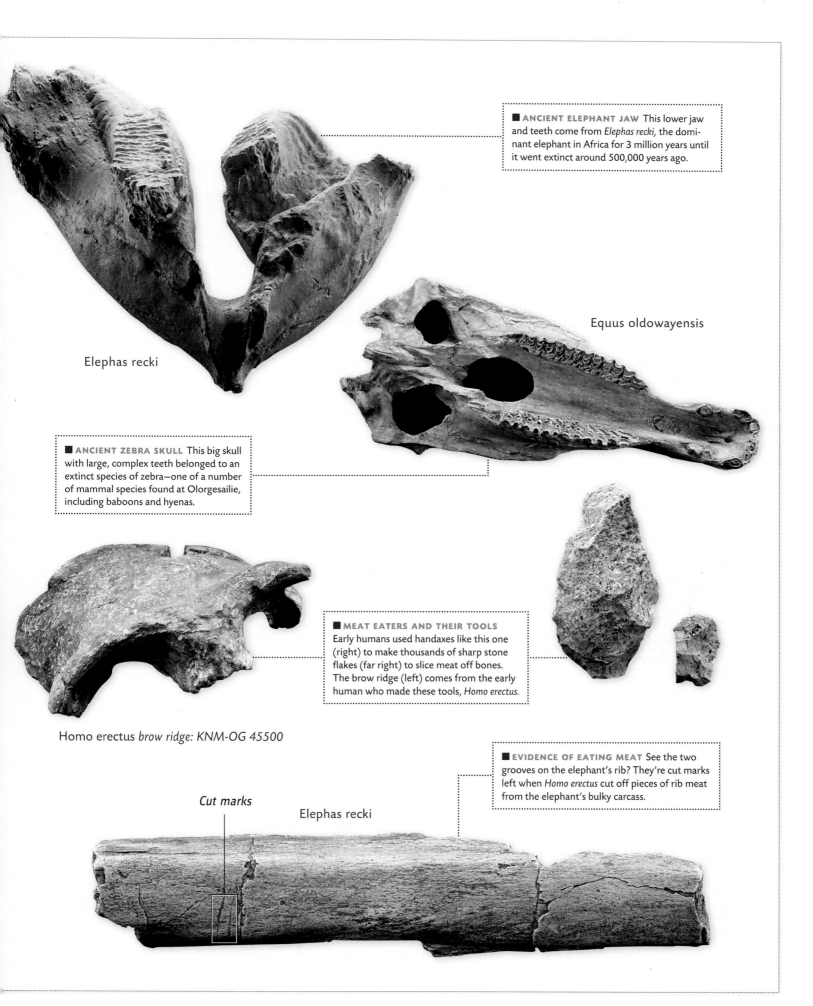

■ ANCIENT ELEPHANT JAW This lower jaw and teeth come from *Elephas recki,* the dominant elephant in Africa for 3 million years until it went extinct around 500,000 years ago.

Equus oldowayensis

Elephas recki

■ ANCIENT ZEBRA SKULL This big skull with large, complex teeth belonged to an extinct species of zebra—one of a number of mammal species found at Olorgesailie, including baboons and hyenas.

■ MEAT EATERS AND THEIR TOOLS Early humans used handaxes like this one (right) to make thousands of sharp stone flakes (far right) to slice meat off bones. The brow ridge (left) comes from the early human who made these tools, *Homo erectus.*

Homo erectus *brow ridge: KNM-OG 45500*

■ EVIDENCE OF EATING MEAT See the two grooves on the elephant's rib? They're cut marks left when *Homo erectus* cut off pieces of rib meat from the elephant's bulky carcass.

Cut marks

Elephas recki

CHAPTER 9

HUMAN INNOVATION

THE PERSISTENCE OVER TIME OF EARLY HUMAN WAYS OF LIFE AND TECHNOLOGIES IS mind-boggling when compared with the pace of change today. The evolution of our capacity to accumulate innovations laid the groundwork for building technologies far beyond a basic toolkit, and for diversifying the possibilities for our species beyond the habits of the handaxe makers.

Two basic patterns of toolmaking—the Oldowan and Acheulean traditions—endured from 2.6 million to 500,000 years ago with hardly anything new added, except for handaxes and other large cutting tools around 1.6 million years ago. These technologies were defined by elementary procedures of flaking stone, which served our hominin ancestors well for a very long time. Then, intriguingly, variety and innovation began to blossom, with a particularly rapid expansion in creativity within the past 100,000 years.

As the handaxe tradition met its demise, technology started to become defined by careful preparation of stone, a wider variety of raw materials, and a smaller and more diverse toolkit. Specialized implements and equipment allowed human ancestors to prepare pigments, process wild grains, store food, and capture fast and dangerous prey. The pace of innovations multiplied exponentially. Rather than lasting for a million years, newer technologies endured for tens of thousands and then thousands of years. Today rapid obsolescence is assumed. We have become entirely dependent on technology for survival, whether we are getting our energy from nuclear reactors or a campfire. Human innovation and reliance on technology are hallmarks of being human.

By around 500,000 years ago, initial hints of innovation begin to appear. A circular wound in the shoulder blade of a butchered horse, found at the site of Boxgrove in England, suggests that a wooden spear had killed the animal. The oldest wooden spears found so far are slightly later, from the site of Schöningen in Germany, dated 400,000 years old. By this date, one of our predecessors, *Homo heidelbergensis,* was hunting big game, such as horses, rhinoceroses, and large deer. No such evidence has yet been found for earlier ancestors.

Around this same time, Acheulean toolmakers began to invest greater time and effort in the craft of making handaxes. Large cutting tools were transformed in appearance from oval stones struck 20 or 30 times to highly refined, symmetrical objects with dozens of carefully placed strikes and an obvious shape in mind. So carefully crafted, the new-style handaxes seem to go beyond what

Previous pages: A geisha views herself in a mirror. The ability to reflect upon oneself and understand the complex minds of others in society is uniquely human.

Opposite: Pipes of a petrochemical plant speak to the complexity of technology today and the rapid pace of innovation.

TODAY

Plant and animal domestication
By 10,500 years ago

Making baskets and pottery
By 26,000-18,000 years ago

Making well-fitted clothing
By 30,000-24,000 years ago

62,500 years ago

Special fishing tools
By 90,000-70,000 years ago

Tools for capturing fast and dangerous prey
By 100,000 years ago

125,000 years ago

187,500 years ago

Communicating with symbols
By 250,000 years ago

250,000 years ago

PAST

was necessary to carry out the job of butchery. As time went on, new tools and strategies for acquiring food developed—a shift to what researchers call modern human behavior, typical of our own species, *Homo sapiens*. That transformation from early human ways of life to modern behavior concerns how we define ourselves. The subject, naturally, has been the center of avid debate.

WHEN AND WHERE?

When anthropologists talk about modern human behavior, and its origins, they are concerned with four dimensions: technological, social, ecological, and cognitive. Regarding technology, modern human behavior involves innovation and the ability to respond to the surroundings in a variety of ways, which are responsible for the diversity of cultures, technologies, and styles of making things. Socially, it involves the ability to form networks of individuals and groups that can exchange information, ideas, and resources. Ecologically, modern human behavior refers to the capacity to use and alter the immediate surroundings, enabling people to disperse to new regions and to use and alter a wider swath of habitats. Each of these dimensions also implies a cognitive acuity that is greater in degree, if not kind, than that of earlier ancestors.

Ideas about when and how these capabilities arose typically fall into two camps—the late/rapid hypothesis versus the early/gradual hypothesis. These labels reflect different understandings about the timing and tempo of the evolution of our behavioral characteristics.

According to the late/rapid viewpoint, an inborn capacity to innovate evolved abruptly between 50,000 and 40,000 years ago and may have resulted from a single mutation that enabled complex symbolic language. The strength of this case rests on the European archeological record, which documents a rapid rise in innovation starting around 40,000 years ago. The key developments include the production of art objects, personal adornments such as beads, sophisticated stone blades, and architectural structures such as huts made from several tons of mammoth bones. The innovations also include the expansion of hunting abilities made possible by projectile weapons and the creation of sophisticated burials and elaborate rituals. Among the most astonishing aspects of this "creative explosion" are the galleries of cave paintings in France and Spain, which reached their height between 32,000 and 18,000 years ago.

The strength of the early/gradual view, by contrast, focuses on the African archeological record before 40,000 years ago. According to this view, a greater variety of stone tools and a faster pace of innovation are evident beginning nearly 300,000 years ago, so the hallmarks of modern human behavior would have arisen gradually.

Differing hypotheses like these can be tested only by looking at the mounting array of discoveries. What were the milestones in the gathering momentum of human innovation? And did they occur over a lengthy period or only near the end of the story?

THE REVOLUTION THAT WASN'T

The most impressive case comes from discoveries in Africa, where an astonishing body of evidence has been amassed by teams of archeologists. A thorough compilation by Sally McBrearty and Alison Brooks, who have excavated for decades in eastern and southern Africa, lends support to the early/gradual hypothesis and has changed the way many researchers think about the transition to modern human behavior. They refer to their conclusions as "the revolution that wasn't," meaning that technological, social, and ecological changes associated with the evolution of our species took time to unfold. It was by no means a revolution in the cognitive capabilities of our species, and the changes may even have started to occur before *Homo sapiens* first emerged.

Well-crafted blades and points, which were once considered diagnostic of the late/rapid revolution, have been discovered in excavations in central Kenya dating to more than 285,000 years ago. Grindstones for the processing of pigments, which could signify the earliest use of color for symbolic purposes, are at least 250,000 old. From the site of Twin Rivers in Zambia, pieces of hematite and limonite, which are the sources of red and yellow ocher, have flattened areas where the pieces were rubbed, indicating that they were held and used like chunky crayons.

The use of ocher, the manufacture of tiny stone blades potentially hafted for use as weapons, and the exploitation of shellfish are all seen at Pinnacle Point in South Africa, and are around 164,000 years old. Shell beads with holes drilled by a standard process and colored by pigment are known from sites in Algeria and Israel back to about 135,000 years ago. The implication is that necklaces, bracelets, or other forms of personal adornment go back much further than anyone had previously thought. The exchange of obsidian among groups over distances of as much as 300 kilometers (186 miles) occurred by about 130,000 years ago,

Opposite: *Hooks and nets, like the one this Austrian fisherman uses, reflect later innovations in fishing; they came after harpoons and spears—among the earliest complex bone tools. The ability to plan for successful fishing went hand in hand with technological innovation.*

Improved tools, such as these long 400,000-year-old spears from Schöningen in Germany, may have permitted early hunters to keep their distance from large prey that could inflict injuries.

and specialized treatment of the dead—based on prepared and polished skulls from the site of Herto in Ethiopia—is dated about 160,000 years old.

A little more than 100,000 years ago, the rate of invention of new fishing and hunting technologies began to rise. Small stone points from Omo Kibish, Ethiopia, dated about 104,000 years old, show evidence that early humans were making composite tools consisting of a point, a shaft, and a natural adhesive or sinew for assembling the parts. The stone points were hafted to make throwing spears, darts, and other projectile weapons, enabling humans to capture fast-moving prey like birds and large, dangerous animals like buffalo and bushpigs. Hunting activities involved planning and strategic positioning of habitations, such as at Porc Epic Cave, Ethiopia, which was situated to overlook areas where migratory herds would pass. At Katanda in the Democratic Republic of Congo, barbed bone points 90,000 to 80,000 years old were found along with the remains of huge catfish, indicating spear or harpoon fishing that could bring in catches large enough to feed dozens of people.

Varying styles of manufacturing points show up in different areas of Africa, which suggests that cultures were beginning to diversify and the behavioral options of our species to multiply. Polished bones 75,000 years old, discovered at Blombos Cave, indicate that early members of our species were already working materials other than stone. Bone awls from this site may have been used to make simple clothing, and nearly a dozen ocher plaques with etched designs appear to reflect some kind of notation or recordkeeping.

In short, there is solid evidence that the suite of technological innovations, social networks, and ecological capacities that characterize *H. sapiens* emerged gradually and were focused in Africa.

The developments after 40,000 years ago in Europe were nonetheless extraordinary. However, they did not evolve rapidly within Europe; rather, they began to occur upon the arrival of populations of *H. sapiens* that had dispersed

The technique for making Acheulean handaxes (this page, left) was widespread across Africa, parts of Asia, and Europe. Between 500,000 and 200,000 years ago, handaxes became smaller and better made (this page, center and right). By 200,000 years ago, handaxes were succeeded by points and blades (opposite page), representing a smaller, more mobile technology.

out of Africa. This legacy includes bone needles essential for making snug-fitting clothing; specialized implements for carving bone, antler, ivory, and wood; spear-throwers for bringing down mammoths and sophisticated harpoons for catching seals; ceramic statues and evidence of a long-lived belief system reflected by the production of Venus figurines; and the efflorescence of cave art. These inspiring innovations denote a zenith in the resourcefulness and imagination of a species that was already reliant on a rich mental and symbolic universe.

That still leaves us with the intriguing question of what was different about our species. What factors contributed to our capacity for innovation, and how did we differ from earlier hominin species?

THE ACCUMULATION OF INNOVATIONS

Leaving the archeological record for the moment, we can focus on what has been learned from studies of our living primate relatives. With patient observation, field researchers have discovered that other species of primates can also be quite inventive. Jane Goodall's work with the Gombe chimpanzees in Tanzania has produced numerous instances of individual chimpanzees taking advantage of accidental happenings to figure out, for example, that a stick could be used to knock fruit to the ground, or that wadded-up leaves mainly used to wipe the body could also help brush away stinging insects. These may seem like simple examples, yet they illustrate how our primate relatives can solve problems in novel ways, use objects creatively, and develop new behaviors that can be learned and passed on over time.

One of the most noted examples of primate insight involved a young female rhesus monkey, named Imo, who began to carry sweet potatoes to a nearby stream to wash off the sand. Shortly after, Imo's mother began to do so, followed by other young individuals in Imo's troop. The technique caught on, and eventually most of the troop washed their sweet potatoes before eating them.

The earliest evidence for fire and cooking

Fire would have provided early humans with light, warmth, and protection from predators at night. Cooking was important because it would have given them access to nutrients in meat and vegetable matter that were otherwise unavailable.

There may be evidence that our ancestors used fire as early as 1.5 million years ago at Swartkrans Cave in South Africa, where scientists uncovered bones burned at temperatures associated with campfires rather than natural fires. Yet there are no campfires or hearths at Swartkrans to indicate that early humans living there controlled fire.

The 790,000-year-old site of Gesher Benot Ya'aqov in Israel is currently the oldest known site showing strong evidence that humans controlled fire to cook food in hearths. Tiny bits of flint, some deformed by heat, cover the site. The concentrations of burned flint, seeds, and wood mark the locations of hearths, while the unburned flints indicate their outer edges.

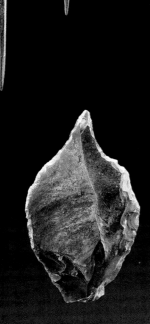

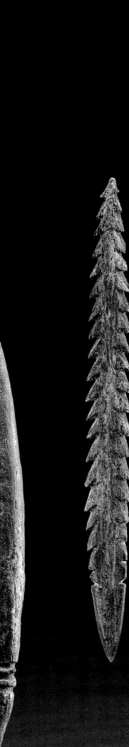

MAKING CLOTHING

Awls and perforators, such as the example from Laugerie Haute, France (above, bottom), were probably invented in Africa and carried to colder climates, where they were used to pierce holes in clothing. Later humans used bone and ivory needles (above, top) to sew warm, closely fitted garments.

FISHING

More than 70,000 years ago, humans in central Africa made some of the earliest barbed points (above, left) to spear huge prehistoric catfish. Later, humans used harpoons (above, right) to hunt large, fast marine mammals.

CARVING AND SHAPING

Burins (above, top and bottom) are specialized stone flakes with sharp, chisel-like tips. Humans used them to work bone, antler, ivory, and wood and to carve designs and images, illustrated by the 17,000-year-old engraved bone from El Pendo Cave, Spain (above, middle).

HUNTING FAST AND DANGEROUS PREY

When attached to spears or darts, stone points, like those from Blombos Cave, South Africa (above, top two), enabled humans to exploit fast-moving or dangerous prey. Later, spear-throwers (above, left) provided leverage for hurling spears and darts greater distances with more speed and accuracy and with less chance of injury from prey.

In this reconstruction, a stone burin is used to make engraved patterns.

To create, adopt, share, and build upon one innovation after another has enabled the growing complexity of human technology and ecological success.

What is interesting about this example is not only Imo's ingenuity but the fact that the technique was copied and spread through the group. It is one thing to be inventive; it is another for the novelty to spread and to persist for any length of time. This is vital to understanding the role of innovation in our own evolutionary history and its possible transforming influence on our species.

The origin of inventiveness partly resides in our primate heritage. Yet the unusual quality of innovation in *Homo sapiens* is not solely a matter of additional cleverness or insight; it is the ability to accumulate innovations that has proved so crucial. To create, adopt, share, and build upon one innovation after another has enabled the growing complexity of human technology and ecological success. The accumulation of innovations was essential to the transition from an ancient hominin way of life, like that of the Acheulean toolmakers, to the creative momentum reflected in the later prehistoric record of *H. sapiens.*

Two factors can make a lot of difference. The first is population density and the connectivity between groups. No matter how inventive certain individuals may be or how easily a novel activity may be adopted, there is little chance for it to persist if it is confined to a small group. Computer models of the ways in which inventions catch on show that a strong network of social contacts is important

FAQ:
What drives innovation?

Beginning at least 2.6 million years ago, early humans used Oldowan stone technology for more than a million years, followed by Acheulean technology for more than another million years. It was only around 100,000 years ago that the pace of technological innovation sped up. Changes in technology are driven by human adaptability and the accumulation of knowledge.

Our large, complex brains can store decades' worth of input and process it in split seconds. Books and computers provide us with an unprecedented volume of the information amassed over many generations. We can respond instantly to events, and we possess physical and social adaptations that enable us to solve problems, which is the essence of innovation.

for a novel behavior to spread widely. Transmitting the innovation across groups is also vital if it is to endure long enough to become part of a growing pool of innovations. One of the controversial aspects of the early/gradual hypothesis of modern human behavior is that, although a series of innovations did indeed occur, each one may have been relatively short-lived. The necessary cognitive and social capabilities seem already to have been in place nearly 300,000 years ago. Yet the innovations evident in the archeological record did not always endure or spread across the continent wherever groups of *H. sapiens* lived.

The second factor concerns environmental change. The Acheulean tradition of toolmaking in East Africa, where it persisted the longest, began to teeter and disappear during a period of strong oscillation between wet and dry climate, beginning around 360,000 years ago. This was the beginning of a prolonged period, roughly 300,000 years in length, when landscapes and their resources were susceptible to large-scale variations. It is during this period that the early indications of human innovation were first expressed. The unchanging, all-purpose handaxe was supplanted by a smaller, more mobile technology. The pace of innovation increased, and the novelties caught on for long enough to be preserved as archeological evidence. Wider social networks and group exchange occurred on occasion, and we see the expression of complex symbolic behavior. This suite of capabilities makes a great deal of sense as a response to environmental change. Mobility, planning, new types of tools, and contact between groups could help reduce the risks and heighten the chances of survival in the most difficult times.

Another effect of environmental change is on population, which, as noted earlier, has important consequences for how readily inventions are adopted. Repeated droughts in Africa between about 140,000 and 70,000 years ago may have reduced population density below a threshold required for the spread of innovations, and prior innovations may even have disappeared. In northern latitudes, the ice ages also had an important influence. As the northern ice sheets grew to their greatest extent, beginning around 33,000 years ago, migrating groups of our own species had already become so competent at surviving difficult times that the population didn't crash. Instead, these early colonists crowded into the most favorable foraging grounds, as reflected by a great number of archeological sites. As population density rose, ideal conditions were created for the spread and accumulation of innovations. The "creative explosion" in Europe may not have been the major milestone in human evolution it was once considered. But it was a real phenomenon indicative of the dual factors—environment and population—that helped fuel the accumulation of innovations.

Eventually, the basic toolkit of ancient hominins gave way to the myriad toolkits and cultural variety of *H. sapiens* as it spread into new environments. Dependence on technology was but one part of a package that made human beings so successful at this time. Yet no element was more important than our imagination, which gave us the ability to contemplate our role in the universe and to plan our future.

Around 18,000 years ago in Japan, the first use of fired clay to make ceramic pots, like this 8-centimeter one, initiated an era of pottery use that continues to this day.

THE ROOTS OF IMAGINATION

THE DORDOGNE REGION OF FRANCE IS MAGICAL. FAIRY-TALE CASTLES AND CHÂTEAUS grace the tops of vertical limestone cliffs carved by a meandering river," wrote Chris Sloan on a *National Geographic* assignment. "Deep within these karst walls our ancestors took refuge in subterranean caverns. It was there, tens of millennia ago, they created the first permanent records of humankind's ability to capture a world within our minds and then release it, transformed, in the form of art. Who were these people? It was not until I witnessed the paintings at Lascaux that I had my answer. As my eyes adjusted to the dim light, enormous images of horses and aurochs loomed overhead, painted on the cave surface with a clear mastery not only of animal anatomy and motion, but the ability to imbue symbols with an awesome and magical power. 'I know who these people were,' I recall saying to myself. 'They're us.' "

Opposite: Intricate patterns adorn the feet and hands of a Yemeni woman. Preparing and enhancing our appearance may be one of the most ancient forms of symbolic communication.

Symbols, the stuff of art, music, language, and ritual, are so integral to our lives that it is hard to imagine life without them. Daily routines like selecting clothing and jewelry to wear, reading the newspaper, chatting with friends, and going to school or work would not exist in a world without symbols. It would be a world with hardly any means to communicate complex ideas and virtually no way to know what is happening beyond the range of your own vision or hearing.

It is hard to overestimate the power of symbols. The Christian cross, Judaism's Star of David, and Islam's crescent moon are but a few symbols around which billions of Earth's inhabitants fervently rally. Flags fly and anthems play as symbols to remind us of our group identity. A color can communicate mood, a piece of jewelry declares marital status, and in our wallets and purses we carry symbols of our wealth in the form of credit cards and cash. Traffic signs, weather maps, newspaper headlines, television, and the Internet all keep us informed through symbols or symbol-based communication.

The emergence of symbolism as a fundamental part of modern human behavior occurred gradually, but we can think of it in terms of three milestones. The earliest known artifacts that probably reflect symbolic behavior, about 300,000

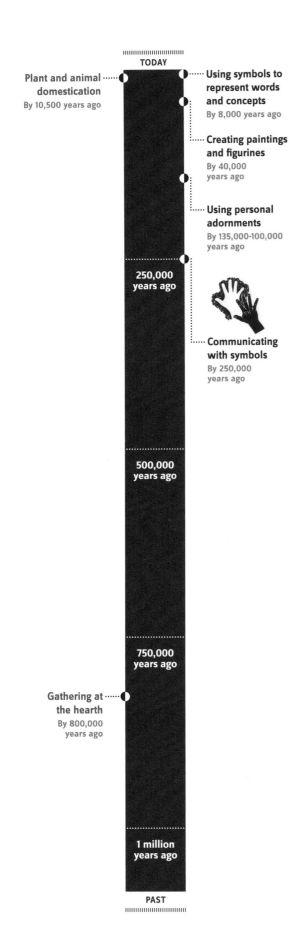

TODAY

Plant and animal domestication
By 10,500 years ago

Using symbols to represent words and concepts
By 8,000 years ago

Creating paintings and figurines
By 40,000 years ago

Using personal adornments
By 135,000-100,000 years ago

250,000 years ago

500,000 years ago

750,000 years ago

Gathering at the hearth
By 800,000 years ago

1 million years ago

Communicating with symbols
By 250,000 years ago

PAST

to 250,000 years old, compose the first milestone. A second milestone can be set at 110,000 to 75,000 years ago, when modern humans increased their use of symbolic behavior; archeological sites from this time period preserve a convincing array of symbolic artifacts. By the third milestone, which can be set at 60,000 to 30,000 years ago, symbolic behavior reached great complexity, as represented by the florescence of cave art in Europe and elsewhere.

SYMBOLIC LANGUAGE

The most powerful symbols of all are those of language. Every human group uses arbitrary symbols and a language for transmitting information and learned behaviors among individuals and across generations. Language enables humans to think about the past and the future, to imagine distant places, and to describe things, such as ideas, that we cannot see. Language is the essential medium through which we share vision, knowledge, meaning, and identity.

Language depends upon agreement among people on what is symbolized by an expressed sound or word. Once the meanings of sounds are agreed upon, they can be combined and recombined to create an open-ended, flexible system of communication. And once visual symbols for words or sounds are agreed on—whether as Chinese logograms or the Phoenician-based alphabet you are reading here—a written language is created.

With symbolic language in place, our ancestors were able to communicate in new ways. Not only could they now reach across time and space but they could also share the secrets of their minds—their hopes, dreams, and memories. They could also delve into abstractions, such as efforts to explain the ways of the world and their very existence. Symbolic communication lies at the root of human imagination.

PRIMATE USE OF SYMBOLS

Are we the only creatures that use symbols? Many animals vocalize, yet there are no known examples of symbol use in the wild. For several decades, though, researchers have run successful efforts to train primates to use symbols in laboratories. Kanzi, a male bonobo born in 1980, has learned to recognize and use 348 images, and he understands more than 3,000 English words. Koko, a female gorilla born in 1971, can communicate using more than 1,000 gestures of American Sign Language.

Primate research also shows that the capacity for symbolic communication through speech is present in species other than humans. In one series of experiments designed to probe the depth of their capacity for language, monkeys were able to distinguish between Japanese and Dutch, no matter who was speaking, and were able to make such distinctions only when the languages were played correctly. This and other research suggests that creatures other than humans, including birds as well as nonhuman primates, are able to discern specific properties of speech, implying that the core mechanisms for speech perception evolved long before speech existed.

So what is unique to human language? First, only we can store a multitude of audible and visual symbols in our brains and then present them in an infinite variety of meaningful combinations. Human language is also recursive, which refers to our ability to wrap meaning within meaning in neat word packages. It permits us to use and comprehend phrases such as "My brother's wife's uncle." Without it one would have to break this phrase into these separate parts: "I have a brother. My brother has a wife. My brother's wife has an uncle." The elegance with which we can communicate complicated plans, conditions (what to do or think if X or Y happens), and social relationships is made possible by this unique feature of language.

How and when vocalizations evolved into language is difficult to know because this transition did not fossilize. One can infer that the evolution of language capabilities was associated with the increasing size and complexity of our brains, compared with other primates'. Despite having an innate ability to learn

Opposite: *Incised with mysterious geometric lines—perhaps early notation—numerous pieces of ocher 100,000 to 75,000 years old were found at Blombos Cave in South Africa.*

Below: *Calligraphic letterforms brushed onto rice paper by a Japanese Zen abbot share features common to both symbolic art and written language.*

Palette

Hematite stick

Prehistoric palette (top), a flat stone from the French cave of Tarté, preserves red pigment from ground ocher used by ancient artists about 20,000 years ago. The flat facet of a hematite stick (bottom) from Zambia represents evidence of humans using orange-red pigment about 250,000 years ago.

Above, right: *"La Clairvoyance," by René Magritte, captures the concept of abstract thinking: the ability to imagine something not immediately perceived by the senses.*

symbols and understand some human language, apes are unable to fully grasp the rules employed to make language work, and it is difficult for them to acquire large vocabularies. The average English-speaking child, who starts talking at around age two, learns about ten words per day. By the end of high school, an average student may have learned about 40,000 words, and by the end of college, perhaps 60,000 to 70,000 of the hundreds of thousands of words in the English language. A chimpanzee learns about 0.1 word a day.

SPEAKING

Key to human language is our ability to make a wide variety of sounds. This is possible through the low position of our larynx, or voice box, compared with its position in the vocal tract of other mammals, including chimpanzees. Combined with our short, round tongue, the low position of our larynx creates a long and wide pharynx, the area at the back of the mouth above the voice box. This expanded pharynx, combined with the action of our lips and tongue, is very effective at rapidly modifying sounds produced by the larynx.

Because it is largely composed of soft tissue, the vocal tract anatomy of our ancestors has not been preserved in fossils, leaving researchers to make inferences from fossils. Recent studies suggest the anatomy required for fully modern speech includes an approximately one-to-one ratio between the lengths of the oral cavity and the vocal tract. This proportion appears to be related to the amount of forward projection of the face. Since the reduction and withdrawal of the face to a position directly beneath the forehead characterizes *Homo sapiens,* including fossil specimens but not other hominin species, earlier species may not have had the ability to control the production of sounds or to organize their rapid production to the degree that characterizes modern human speech and language capabilities.

Modern human language capabilities required more than an evolved vocal tract. Brain structures that regulate motor control and cognitive processes had to evolve as well and were most likely part of the evolving structure and wiring of the human brain over time. Another line of evidence for the evolution of these structures can be found in our genes. FOXP2, one regulatory gene that contributes to the development of brain structures associated with speech, language, cognition, and motor capabilities, is different in humans and chimpanzees by two mutations. Using molecular genetics, one study estimated that the human form appeared within the past 200,000 years. A later study of genetic material obtained from Neanderthal bones showed, however, that Neanderthals also possessed the human form of the FOXP2 gene, and that this genetic factor, probably one of many associated with language production, may have occurred in a Neanderthal–modern human ancestor by about 500,000 years ago.

Our wonderful apparatus for speech comes at one great cost. Because of the low position of the larynx, we cannot completely close off our air passages while we eat. We are the only mammal that cannot breathe and swallow at the same time. As a result we are constantly at risk of choking. Through an amazing evolutionary balancing act, our infants are born with the larynx in a high position, which permits them to suckle, swallow, and breathe at the same time. Only after several months does the larynx begin to descend slowly, eventually imparting to the child an ability to speak when the language functions of the brain are mature enough to do so.

THE FIRST GLIMMERS

Language and symbolic behavior added an entirely new dimension to our ancestors' existence. Symbolic thought allowed us to harness the power of consciousness—one of humanity's most prominent defining qualities. Language provides a way of processing mental images and assigning meaning to objects, events, and abstractions that need not be visible. An amazing aspect of language is its ability to process sensations about the mind itself and to locate mental activity where it actually occurs, inside each one of us. We can symbolically identify and distinguish our own mental life and personal experiences from those of others. We can imagine and talk about "my mind" and "your thoughts." We can think about thinking and our own identities. Thinking, emotions, and the whole spectrum of mental experiences can be understood and used as tools that serve one's own concerns and those of others, conferring an immense survival benefit.

Marine snail shells, each perforated to fit tightly in a necklace, grace a reconstructed necklace from Cro-Magnon, France, about 30,000 years old. Such worked shells represent some of the earliest evidence of humans wearing jewelry.

Laid head to head when they died about 24,000 years ago at Sunghir in Russia, the bodies of two children were adorned with beads and ocher, possibly signs of special social status.

While we cannot know what our ancestors were saying to each other, we are able to glimpse what a prehistoric world charged with symbolism may have been like. Symbols became so important that they entered the realm of the things that our ancestors made and left behind, providing a whole new line of evidence for the origins of symbolic behavior.

Some researchers see the first glimmerings of symbolic behavior more than two million years ago; others disagree. However, there is enough evidence to suggest that while not common, certain symbolic behaviors were not anomalous by 300,000 to 250,000 years ago. Among these activities were the use of pigment and the altering of natural objects.

Of the two, the use of pigment is more commonly encountered by archeologists. Pigments can be made from many natural substances, such as carbon and manganese for black, and kaolin for white, but the most common pigment-associated mineral is ocher, or iron oxide. This can take the form of red ocher (red hematite) or yellow ocher (yellow limonite). One can tell from the way some ocher pieces are worn that they were being used as crayons or being ground into powder; when made wet, the brightly colored pigment could be applied to almost any surface, including the human body. Though there is little evidence, it's possible that much of the earliest use of ocher was on perishable material, such as wood and human flesh.

An example of an altered natural object from this early period is the Berekhat Ram figurine from the Golan Heights. Ancient artists apparently imagined the female form within the natural shape of this stone and emphasized it by artificially deepening the natural grooves. The stone, which was carved 280,000 to 250,000 years ago, may have been covered in red ocher.

ADORNMENT AND BURIAL

The simplicity of these early examples of human symbolic behavior contrasts with the more complex examples made by around 100,000 years ago, when *Homo sapiens* was present. Evidence of symbolic behaviors at this milestone comes not only from the use of pigment but also from the crafting of necklaces and pendants and the earliest mortuary rituals.

Modern humans made beads from more than 150 different materials. These included animal and human teeth, bone, eggshell, and seashells. It is clear these objects were highly valued and carefully crafted. The fine circular shape and consistency in size of tiny ostrich eggshell beads from El Greifa in Libya, which may be as much as 200,000 years old, suggest they were the product of a standardized manufacturing method, as were numerous *Nassarius* gastropod shells, destined for necklaces, found at several Mediterranean coastal sites.

Both ocher and prehistoric jewelry are often found in association with early burials, another hallmark of human symbolic behavior. Some of the earliest known human burial sites, at Skhūl Cave and the Jebel Qafzeh rock shelter in Israel, are between 135,000 and 90,000 years old. Bits of ocher were associated with burials at Skhūl, as were shell beads and a large boar mandible. Shells and ocher were also present at nearby Jebel Qafzeh, and one individual may have been buried with a grave offering of fallow deer antlers.

Burial was not the only symbolic mortuary ritual of our ancestors. Defleshing and cremation were also practiced. Stone tool marks on 160,000- to 154,000-year-old skulls from Herto in Ethiopia suggest these heads were removed from bodies and then defleshed carefully. The polished surface on the skull of a child from Herto suggests that it may have been carried for some time, possibly as a ritual object. The earliest known humans in Australia were also found in a symbolic burial context, cremated and sprinkled with ocher. This shows that humans were carrying their symbolic behavior with them as they spread around the globe.

Around this same time, earlier members of our species also nurtured ideas about counting and keeping track, as indicated by gridlike patterns incised on bone, ivory, and stone. Some of the earliest examples, from Blombos Cave in South Africa, involved patterns of intersecting diagonal lines etched on 78,000-year-old pieces of red ocher.

Following pages: *Ancestor of today's cattle, a prehistoric aurochs dominates a wall painted around 17,000 years ago at Lascaux Cave in France.*

The first burial rituals

Neanderthals and modern humans are the only hominins known to intentionally bury their dead. But the clearest and least controversial ceremonial burials are those currently associated with modern humans. The earliest of these at Qafzeh, Israel, is more than 100,000 years old, and it includes red ocher and ocher-stained stone tools—indications of symbolic behavior. Later burials, such as one of a boy and a girl at Sunghir in Russia 24,000 years ago, show some of the first signs of social status. These two youths, perhaps the children of a chief, were buried head to head, flanked by two huge mammoth tusks, covered in ocher, and ornamented with beaded caps, carved pendants, and thousands of ivory beads.

ART

On December 18, 1994, cave explorer Jean-Marie Chauvet and two friends probed the pitch-black recesses of an underground cavern in France. Suddenly, the beam of light from the lamp of one of Chauvet's companions swept across a drawing of a small mammoth prepared in red ocher. "They were here!" she called out.

Further careful exploration of the cave, which came to be known as the Grotte Chauvet, revealed hundreds of paintings and engravings left by humans who had lived there as long as 32,000 years ago. In this cave, prehistoric artists, perhaps shamans, used a wide selection of pigments, which, when mixed with water, fat, or egg, provided a palette as broad as their imagination. Most animals they depicted were herd animals, such as horses, reindeer, and aurochs. Many were shown wounded or pregnant. On other walls, they painted prides of lions, bears, and hyenas—animals that not only competed with humans for prey but also sometimes treated humans as prey.

Why were our ancestors drawn deep within caves like this? Was it just for shelter, or was it for initiation rites, fertility ceremonies, and rituals designed to bring luck to hunters? Some researchers subscribe to the latter idea, suggesting that this would have been particularly important from around 18,000 to 10,000 years ago, as the last ice age was winding down and herds of game animals were dying out or moving away.

Whatever their symbolic function, these sites were important to whole groups, not just isolated individuals. Along with adult footprints, many footprints of children and adolescents are preserved in cave floor sediments. A recent study of ocher hand stencils concluded that many of the hands were those of women.

Illustration of human and other figures may have originated in Africa, where a site called Apollo 11 Cave in Namibia preserves perhaps the oldest known drawing of a human figure, in charcoal on a stone slab, between 60,000 and 40,000 years old. The florescence of art by 32,000 years ago was not restricted to paintings and engravings on stone plaques or cave walls. Numerous carvings using mammoth bone, ivory, and antler also began to appear at the same time.

> **A figure of a fantastic creature, a man with a lion's head, indicates a capacity for imagining figures that exist in the mind but are not of the observable world.**

FAQ:
Did Neanderthals have symbolic behavior?

While fossil and archeological evidence shows that Neanderthals were accomplished big-game hunters and skilled toolmakers, evidence for symbolic behavior in our closest fossil relatives is less clear. A complicating factor is that Neanderthals and modern humans coexisted for thousands of years and sometimes inhabited the same caves. There is strong evidence, however, that some Neanderthals buried their dead, used pigments, carved enigmatic objects, made jewelry, and may have had basic language skills. However, scientists who study Neanderthals vigorously debate the extent to which these behaviors involved symbolic thought and whether all Neanderthals had these capacities.

The oldest known figurative art in Europe is a 35,000-year-old statue of a female from Hohle Fels Cave in southern Germany. Like the famous Venus figurines that are about 10,000 years younger, this figure has exaggerated breasts and genitalia, suggesting it may be a fertility symbol. A figure of a fantastic creature, a man with a lion's head, was found in slightly younger layers in the same cave and indicates a capacity for imagining figures that exist in the mind but are not of the observable world.

Within the caves of Europe 35,000 years ago, one might also have heard the sound of music. Melodies, and perhaps chants and songs, may have accompanied group activities during rituals or dances. This is speculation, but perhaps it is not going too far. The lion-man and the female fertility figurine were found within the same ancient layers as the world's oldest known instruments, flutes made of mammoth and swan bones.

SYMBOLIC UNIVERSE

The music we play, the language we use, and the art we make, now as then, are all symbolic behaviors that help establish group identity. Using symbols to communicate social status and group identity was probably one of the original purposes of personal adornment. Without saying a word, one can communicate simple ideas, such as "I'm already married," "I'm the chief," or "We're a team."

Today, this type of symbolism is often directed at strangers or people outside the immediate social group. Perhaps personal adornment for group identity or status evolved when our ancestors communicated at a distance or when nonverbal communication offered them a distinct advantage. This would have occurred when populations in Africa reached a large enough size to make such encounters fairly common. Some researchers suggest that regional differences in personal adornment may reflect linguistic differences. The same may be true of other sorts of symbolic objects, such as the Venus figurines and lion-men.

In contrast to the first major dispersal of *Homo,* typically ascribed to *Homo erectus,* symbolic behavior was critical to the success of the dispersal of *Homo sapiens* within and beyond Africa. The ability of people to create and reinforce group identity permitted humans to diversify culturally and to adapt their ways of life—and their identities—to the new conditions they encountered. Humans had developed a means of creating a cultural, symbolic universe that reflected survival conditions in the immediate present but was also full of other possibilities. We were a species ready to apply its gifts for imagining and innovating, which united us in the human condition and allowed our dispersal to all corners of the planet.

One of the oldest pieces of art discovered in Europe, this 35,000-year-old ivory lion-human chimera hints that a rich symbolic life existed in early members of our species.

ONE SPECIES WORLDWIDE

AS EARTH'S POPULATION APPROACHES SEVEN BILLION PEOPLE, IT'S ASTOUNDING TO think that *Homo sapiens* once came close to extinction. The genetic similarity among all people implies that a rare and dangerous interlude occurred recently, in a blink of evolutionary time, when a census of our direct ancestors would have counted people in the mere thousands, not millions or billions. Today, we are one species—a species of many cultures, diverse languages, and individuals of varied opinions but a single lineage nonetheless, with six million years of shared evolutionary experience.

In the 1980s, biochemist Allan C. Wilson of the University of California, Berkeley, achieved international recognition after he and his students found a molecular archive of *Homo sapiens* in the DNA of the cell's nucleus and mitochondria, which every person has inherited from his or her parents back through time. Laboratories at Berkeley and elsewhere began comparing individuals' DNA signatures, which can provide a record of the relationship among everyone around the globe. Geneticists sought answers to questions the fossils could not give about the evolution of our species: When did *Homo sapiens* first evolve? And where did that transition take place?

For some years, it seemed that the genes and the fossils would tell contradictory stories. Since the early 1960s, molecular comparisons among humans and other living primates hinted at a different age for the divergence of humans from our common ape ancestor compared with what the fossils indicated. The debates were tense. The fossil experts considered it *their* job, not the geneticists', to find the branching point where the hominin lineage began. In the 1980s, the geneticists stepped again into the domain of the paleontologists, this time using the genetic diversity of living people to trace the root of our particular species and its recent history.

In the end, these two rich lines of evidence—the genes and the fossils—uncovered areas of strong agreement, and together would become the source of new evidence and ideas about the human journey. These same lines of evidence would produce similar dates not only for the divergence of hominins from an African ape common ancestor but also for the colonization of our species around the globe, and they would show the extent to which people everywhere are alike or different.

Opposite: *A New Year's celebration in Times Square, New York, brings thousands of people together. Millions more join in around the world through TV and the Internet.*

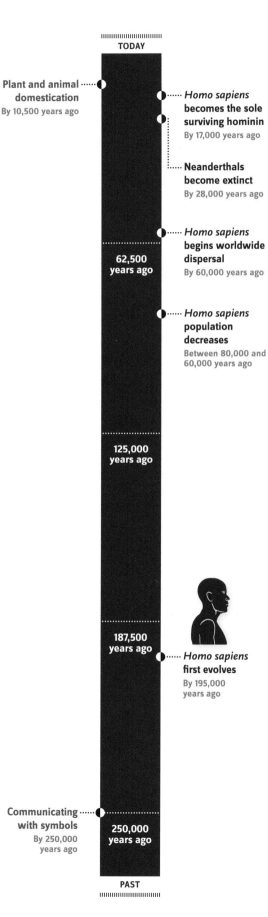

TODAY

Plant and animal
domestication
By 10,500 years ago

Homo sapiens
**becomes the sole
surviving hominin**
By 17,000 years ago

**Neanderthals
become extinct**
By 28,000 years ago

Homo sapiens
**begins worldwide
dispersal**
By 60,000 years ago

62,500
years ago

Homo sapiens
**population
decreases**
Between 80,000 and
60,000 years ago

125,000
years ago

187,500
years ago

Homo sapiens
first evolves
By 195,000
years ago

Communicating
with symbols
By 250,000
years ago

250,000
years ago

PAST

While the results of the DNA analyses have not gone unchallenged, genetic information offers invaluable, independent ways of investigating the antiquity of *H. sapiens,* the most likely point of origin, and the relationship among the world's populations. The clues come from two main DNA sources: mitochondrial DNA and nuclear DNA, with the latter further divided into the informative Y-chromosome DNA and the remainder from the total of 46 chromosomes in humans.

Mitochondria are tiny structures in the cytoplasm of our cells that generate most of the cell's supply of chemical energy. Mitochondria also have their own DNA, abbreviated mtDNA, which is separate from the DNA in the cell's nucleus and can be passed on only as part of the egg cell's cytoplasm. As a result, mtDNA is inherited exclusively from our mother's side. Individuals with identical mtDNA are related to each other and to a common ancestral female, whether a mother, a grandmother, or a more distant female relative.

Like nuclear DNA, mtDNA undergoes mutation, which is responsible for subtle variations that appear in all living humans. These variations are markers that help trace the history of changes in the mtDNA back to the original source—that is, a hypothetical woman who had the original mtDNA mutation from which all subsequent variations developed. Although this original source is usually nicknamed "mitochondrial Eve," mtDNA provides a record of only one of many thousands of genetic histories that came from other original women and men—a wide array of individuals who lived in different places and at different times. Because its rate of mutation is rapid, though, mtDNA is especially useful in tracing the history of human populations and provides some of the most useful estimates of our species' genetic history. Molecular clock estimates, which are based on the rate of mutation, indicate that the original mtDNA source—mitochondrial Eve—lived about 220,000 to 140,000 years ago.

Nuclear DNA is different in that it is inherited from both mother and father, and the recombining of the parents' DNA at conception means that nuclear DNA offers a far more convoluted genetic history of males and females back through time, which is difficult to untangle and interpret. However, the Y chromosome contains DNA that is passed only from father to son. Comparison of the Y-chromosome variations from living humans shows that the original mutation source—the so-called Y-chromosome Adam—occurred between 200,000 and 60,000 years ago. Even though the ranges of dates for Y-chromosome Adam and mitochondrial Eve overlap to some degree, it's likely that the individuals who carried these original DNA variations were separated by many millennia. Not only were they not a couple, these were not the only modern humans who contributed DNA to our genome.

Over the past two decades, an enormous amount of nuclear and mitochondrial DNA has been collected from diverse human populations. The combined genetic information shows that *H. sapiens* is a young species, with an average estimate of about 200,000 years ago for the origin of our species. The DNA data also indicate that African populations have by far the largest amount of genetic diversity. This pattern is what one would expect if our genetic origins were in Africa and small populations left the continent and colonized the world. Specific segments of DNA variation, called haplotypes, can also be identified in living people. These haplotypes can be traced geographically, and they also indicate Africa as the source area of the most ancient DNA variations. Altogether, the DNA clues inside our bodies point to a remarkable conclusion: All modern human genetic variation can be traced to Africa, and only a little piece of African genetic variation is the raw material from which the diversity in non-African humans has evolved.

WITHIN AND BEYOND AFRICA

The age of the oldest known fossils of *Homo sapiens* is in agreement with the estimated time for the origin of our species based on DNA evidence. These fossils show clear differences from our immediate predecessor, *Homo heidelbergensis,* which include a more spherical braincase and a smaller face that is positioned beneath a high forehead. The Omo River region of southern Ethiopia is where these early *H. sapiens* specimens were found through the work of a team led by Richard Leakey, the son of paleoanthropologists Louis and Mary Leakey. Upon the discovery of the fossils in 1967, their initial assessment indicated that these ancestors lived 130,000 years ago. Subsequent visits to the site between 1999 and 2003 by another team found more fossil material and permitted a new analysis of their age. The Omo *H. sapiens* remains are now placed at around 195,000 years old.

At present the fossil record is too sparse to help us understand the expansion of our species within and beyond Africa after this initial date. However, the dual lines of evidence—genetic and fossil—can come together to help reconstruct an outline of the spread of *H. sapiens* around the world.

The breadth of African genetic diversity suggests that there was a complex series of population movements within Africa following the origin of our species. The dispersal of human populations to other continents came later, beginning at least 60,000 years ago. While the genetic diversity of populations beyond Africa is relatively small, it underwent an expansion roughly 60,000 to 40,000 years ago—which reflects the major expansion to other parts of the world.

One complicating factor is that fossils of *H. sapiens* are known back to slightly earlier than 100,000 years ago in the Levant region of the Near East—from the sites of Skhūl and Jebel Qafzeh, Israel. Fossils attributed to our species have also been discovered in southern China at sites such as Liujiang, with an estimated age of at least 68,000 years old. The Levant and southern China fossils may represent part of an initial, short-lived wave of dispersal, whereas the later expansion of African populations to other parts of the world, starting around 60,000 years

Three early **Homo sapiens** *skulls. Despite the diverse appearance, all skulls of our species are characterized by a rounded braincase and the face positioned beneath the brow and forehead.*

Irhoud 1, Morocco
About 160,000 years old

Liujiang Cave, China
At least 68,000 years old

Kow Swamp 1, Australia
About 13,000–9,000 years old

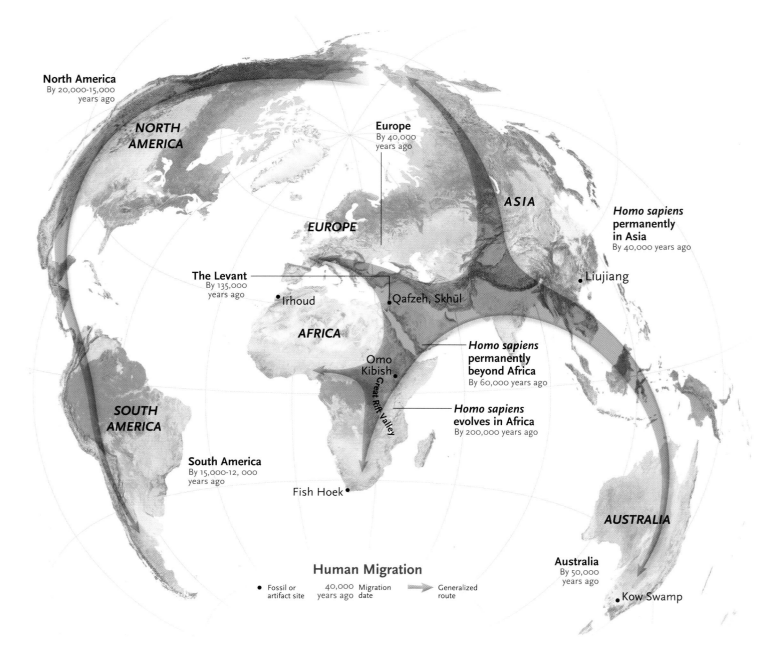

North America
By 20,000-15,000
years ago

NORTH AMERICA

Europe
By 40,000
years ago

ASIA

Homo sapiens
**permanently
in Asia**
By 40,000 years ago

EUROPE

The Levant
By 135,000
years ago

•Liujiang

•Irhoud

•Qafzeh, Skhūl

AFRICA

Homo sapiens
**permanently
beyond Africa**
By 60,000 years ago

Omo
Kibish•

Great Rift Valley

Homo sapiens
evolves in Africa
By 200,000 years ago

SOUTH AMERICA

South America
By 15,000-12,000
years ago

Fish Hoek•

AUSTRALIA

Human Migration

• Fossil or
artifact site

40,000 Migration
years ago date

Generalized
route

Australia
By 50,000
years ago

•Kow Swamp

From our African origins, the migrations of modern humans to other continents can be tracked through genetic, fossil, and archeological evidence.

ago, was so robust that it swamped the genetic signal of the earlier wave. Some interpretations of the genetic evidence posit a complete replacement of older populations by the later expansion, while other models allow for some degree of interbreeding between the colonizers and the more ancient populations in a given region.

MORE ALIKE THAN UNLIKE

It wasn't too long ago that scientific opinion was shaped more by public biases than vice versa. People accepted the notion that humans from different places represented separate species or, even as part of a single species, that different races had separate evolutionary histories. Since 1953, when James Watson and Francis Crick identified the structure of DNA, the science of genetics has become racism's enemy. Studying the total array of genes, scientists in the field known as genomics have eliminated the notion of separate human species or subspecies. There is only one species of modern human: *Homo sapiens.* Genomics has also shown that despite the obvious physical differences among us, we are all far more alike than we are different.

What little genetic diversity we have declines the farther one gets from Africa. Of the variation that exists, about 88 to 94 percent of it is found among individuals who live within the same population, which is often defined, more or less, by geography and language. The difference among populations living on different continents amounts to between 6 and 9 percent of the total genetic variation within our species. Thus the physical features that people often use to distinguish racial groups are likely to have evolved only very recently in comparison with the long evolutionary history shared by all human beings.

Anthropologists and geneticists also see modern human variation as a global web of small gradations from one population to another. Sharp genetic divisions between peoples living in adjacent regions are rare; instead, the geographic variations are gradual, with genetic buffer zones where admixture between populations occurs. This pattern is seen all over the world and is why the division of *H. sapiens* into races is unfounded from a genetic perspective. Scientists thus do not see race as a biological concept but as a learned social concept.

A SPECIES IN TROUBLE

Although humans seem to vary considerably in their physical appearance, our species is actually very homogeneous from a genetic standpoint. Although chimpanzees have a much smaller population and geographic distribution than we do, they vary a lot more genetically than do humans, both in their mitochondrial and nuclear DNA. The low genetic variation in humans means one of two things: All living human populations evolved from an ancestral population that was either very small for a very long period of time or that was large initially but then underwent a dramatic and recent constriction, a process known as a genetic bottleneck. Based on rates of DNA mutation, it is possible to estimate the timing of different genetic mutations, as we have noted for mitochondrial Eve and Y-chromosome Adam, and also to calculate the size of the adult breeding population.

Although there is a range of scientific opinion about the two possibilities, several DNA analyses indicate that modern human genetic variation, sampled from all over the world, goes back to a series of original mutation dates within the past 100,000 years. That would favor the idea of a reduction in *Homo sapiens* population size at a recent time in human evolution. Y-chromosome data indicate that this population reduction, and a genetic bottleneck when many adults died or were unable to reproduce for some reason, took place sometime between 90,000 and 60,000 years ago.

There are at least three scenarios: a decline in the number of reproductively active adults to about 10,000; a prolonged period when as few as 2,000 adults

Who were Genetic Adam and Eve?

For any part of our DNA, there is a genetic ancestor. The human genome is a mosaic of genetic material that has taken shape over billions of years. By comparing the genetic material of many different organisms, researchers have found that each human gene has its own particular evolutionary history. Some genetic variations arose within our own species, others in our primate common ancestors, and still others in more ancient ancestors. "Genetic Adam and Eve" refers to genetic variations that originated in particular individuals of our own species.

Unlike the biblical pair, Genetic Adam and Eve lived thousands of years apart and never would have met. "Y-chromosome Adam" and "mitochondrial Eve" are nicknames given to the hypothetical individuals whose DNA, from part of the Y chromosome and the cell's mitochondria, respectively, represents a common ancestor for the DNA of all modern humans today. Genetic Adam and Eve did not live alone, but among hundreds or thousands of other modern humans, many of whom also contributed to our genetic makeup.

made up the entire breeding population; or, based on combining the data for 50 distinct genetic markers, a sharp reduction to as few as 600 adults for the entire species—or at least for the population that gave rise to all living people today. Each of these estimates, and especially the last two, implies that our species teetered on the edge of extinction, most likely within the past 100,000 years, and possibly as recently as 70,000 to 60,000 years ago. As our species currently approaches seven billion people, it is perhaps hard to imagine that we could ever have been so vulnerable.

A DNA sample in a vial on ice taken from the Neanderthals of Vindija Cave in Croatia has helped unravel the genetic code of our close cousins and confirm that they were not members of our species.

The reason for this die-off is a subject ripe for scientific debate. Natural catastrophes have been invoked as possible culprits in our near demise. One idea implicates the enormous eruption of Mount Toba, in Indonesia, around 74,000 years ago. The catastrophic explosion of ash and rock covered parts of Southeast Asia with many meters of volcanic material, and temperature decreases as far away as Europe may have coincided with the eruption. The effect of the Toba eruption, however, has yet to be demonstrated in the environmental record of Africa, where the genetic bottleneck is thought to have occurred. Another factor could be the extreme climatic variability that wracked parts of Africa between 140,000 and 70,000 years ago. This period included two episodes of severe desertification during which the large Rift Valley lakes almost dried up. Water volume was reduced in Lake Malawi by 95 percent. Vegetation around the lakes essentially disappeared. Only after 70,000 years of dramatic swings in climate did stable, wetter conditions finally return.

The erratic climate and intense aridity in tropical Africa during this period must have stressed the humans living there, forcing them to find new habitat as best they could. The end of aridity in tropical Africa coincided with the appearance of aridity elsewhere in Africa and in the Levant. Arid conditions shifted north from equatorial areas between 90,000 and 70,000 years ago. Researchers suggest that as this shift occurred, it presented a window of opportunity for migration out of Africa. And wetter, more stable conditions, which arrived by about 60,000 years ago, could have favored the rapid expansion of the remaining human populations.

FAQ:
How do we know our species originated in Africa?

Both fossil and genetic evidence strongly supports an African origin for modern humans, about 200,000 years ago. This suggests that modern humans have lived in Africa longer than they have anywhere else; the earliest known modern human fossils are from Ethiopia and are about 195,000 years old. The earliest dates for modern human fossils outside of Africa are almost 100,000 years younger. Living African populations have the largest genetic diversity among humans; genetic diversity decreases the farther populations are from Africa. All non-African populations have a subset of African genetic diversity, indicating that they can trace their ancestry back to a single African population.

THE SURVIVOR

Although we are a relatively young species, the time period of *Homo sapiens* has overlapped with at least three other hominin species. The last populations of *Homo erectus* appear to have been confined to Southeast Asia, with fossils from Java dated possibly as late as 70,000 years ago. *Homo floresiensis* lived to the east on the island of Flores until as recently as 17,000 years ago. The lineage of early humans that has drawn the greatest attention is the Neanderthals. As our own species was declining temporarily in Africa, *Homo neanderthalensis* was thriving in Europe and in western and central Asia. The recent discovery of ancient DNA preserved in Neanderthal bones is now providing human origins research a fascinating new direction. The Neanderthal Genome Project, initiated by the Max Planck Institute for Evolutionary Anthropology in Leipzig, Germany, will ultimately allow scientists to map the genetic information of the Neanderthal line and examine how it differed from *H. sapiens* at the genetic level. Comparisons of Neanderthal and modern human DNA suggest that the two lineages diverged from a common ancestor, most likely *Homo heidelbergensis,* sometime between 350,000 and 400,000 years ago—with the European branch leading to Neanderthals and the African branch to us.

The bigger question is why the Neanderthals became extinct. Some decades ago, researchers assumed they must have met a violent death at the hands of *H. sapiens* who arrived in Europe armed with all the cultural capacities of our species. However, there is no evidence of warfare between the species; in fact,

Homo floresiensis, *left,* and **Homo neanderthalensis** *both survived until relatively recent times. Neanderthals disappeared from Europe around 28,000 years ago, and the last "hobbit" lived to nearly 17,000 years ago on an Indonesian island.*

the oldest evidence of death by a sharp instrument is a severely cut rib, from the thrust of a sharp stone tool into the chest cavity of a 45,000-year-old Neanderthal from Shanidar Cave in northern Iraq. But this is an isolated case; no other similarly injured fossil bones have been found from around the time when Neanderthals went extinct, by about 30,000 to 28,000 years ago.

The Neanderthals possessed a diverse toolkit and made effective use of local stone resources in fashioning tools. They were efficient hunters, although largely reliant on close-quarter weapons such as thrusting spears. The diet of the European-based populations was focused on meat; thus they may have depended on a less diverse resource base than hominin species who lived in warmer conditions. Although their muscular bodies were adapted to the cold, their known distribution remained south of 55° N latitude, from southern England through much of continental Europe, with the Near East as the southernmost part of their range. They practiced burials, although usually with minimal grave goods; and their use of pigments indicates a symbolic capacity.

The toolkit and technological capacity of *H. sapiens* during its movement into the Near East, around 100,000 years ago, closely matched that of the Neanderthals. In fact, modern humans and Neanderthals may have made excursions into the Near East at different times, depending on whether the climate was cold or warm. However, the population expansion that began in Africa around 60,000 years ago, and led to the dispersal of our species to as far as Australia by 50,000 years ago and to the Americas by 30,000 to 15,000 years ago, also led our species to Europe—into regions previously occupied only by the Neanderthals—between about 41,000 and 38,000 years ago. Those European colonizers had behavioral capacities that differed from the

Washington, D.C., children demonstrate wide, continuous variation in skin color, a trait determined by a well-known evolutionary response to the sun's ultraviolet radiation.

Neanderthals'; they used specialized tools and equipment that enabled them to make snug-fitting clothing, build permanent structures, and adjust to varied environments through cultural means. Even though skeletal remains indicate that they possessed slightly elongated extremities typical of warm-adapted bodies, their northern expansion exceeded the limits of the Neanderthal range. Their shell beads and stone material, exchanged over greater distances, suggested the importance of broad social networks among groups. They eventually were responsible for the creative explosion of cave art, expressive of the richness of their cultural and social lives.

It's likely that climate also played a role. The arrival of *H. sapiens* in western Europe, between 38,000 and 34,000 years ago, coincided with a prolonged cold period when the Neanderthals may have moved to the southern regions of Iberia and Gibraltar. By the time warmer conditions prevailed, populations of our species were already ensconced in the regions where Neanderthals would have repopulated. But this time, the behavioral capacities of our species provided a competitive edge. The Neanderthals ultimately died out shortly after 30,000 years ago.

By 17,000 years ago, *Homo sapiens* was the sole survivor of what had been a diverse family tree. Our species had reached all continents except Antarctica. We had encountered every imaginable climate and habitat. Presumably, by that time we had also diverged in our physical appearance after having spent thousands of years in such different places, sometimes isolated from one another. Also by that time, the world was coming out of the harsh grip of the last major ice age. Glaciers were receding, and the ocean carved new shorelines as low-lying land disappeared underwater. A whole new world was opening up as the Earth thawed. And we were the only humans left to inherit it.

By 17,000 years ago, *Homo sapiens* was the sole survivor of what had been a diverse family tree.

LIFE AND DEATH OF A NEANDERTHAL

Neanderthals once frequented a mountainous, cave-pocked region in northeastern Iraq. One cave, Shanidar, was excavated in the 1950s and 1960s by a joint team from the Smithsonian Institution and the Iraqi Department of Antiquities. The excavations produced some of the most remarkable, and controversial, evidence of Neanderthal behavior ever found.

Hundreds of stone tools and butchered animal bones, as well as concentrations of ash indicative of hearths, came to light. The Neanderthal bones themselves exhibit a variety of injuries and illness. Scientists sometimes saw post-injury healing over many years, suggesting the group cared for the ill and injured.

The Shanidar Neanderthals buried some of their dead in shallow graves among the blocks of limestone that had fallen from the cave ceiling. The most extraordinary grave is the controversial "flower burial." In this grave, which is the only one of its kind known to be associated with Neanderthals, the corpse of an elderly man was apparently placed on a bed of pine and fir branches, accompanied by seven kinds of colorful flowers. The pollen and even the anthers of certain flowers were found, and they provide the evidence we need to know the season of the burial: spring.

Although some scientists have suggested that the flower remains were carried in by birds or rodents, among other possibilities, the burial was deep within Shanidar Cave, making these other options unlikely. If the burial did occur with pine, fir, and flowers, it would be a very early and clear expression of symbolic behavior in association with a Neanderthal burial.

Neanderthals add vegetation to a burial site, suggesting they cared for their own, even in death.
They aided the living, sick, and injured, we know with certainty.

■ **STONE SCRAPING TOOL** Neanderthals used stone scrapers like this one to clean animal hides, among other purposes. Researchers found more than 670 stone tools in the oldest levels of Shanidar Cave—evidence that early humans occasionally visited there.

■ **POSSIBLE BURIAL CEREMONY** Soil from this burial site contained pollen from flowers that never could have grown inside the cave. While pollen could have been carried in by rodents or by accident during excavation, there are indications that flowers and branches were brought here for a special purpose.

Homo neanderthalensis
Hand; Shanidar 4

■ **NEANDERTHAL BONES** These are the fossil bones of the elderly Neanderthal; scientists know this because of extensive bone loss in the jaw, caused by teeth falling out. The hand of this same individual is the most complete Neanderthal hand ever found.

Homo neanderthalensis
Mandible; Shanidar 4

CHAPTER 12

THE TURNING POINT

IN THE PAST 10,000 YEARS, HUMANS HAVE TRANSFORMED THEIR SURROUNDINGS BY burning, planting, watering, raising crops, and tending herds. A new way of life emerged, focused on agriculture and cities. One of the basic adaptations of our hunter-gatherer past was sacrificed—mobility, which allowed our ancestors to move wherever food could be found and away from the wastes of their own activities. The trade-off was that domesticated cereal grains offered a new source of protein and energy, and they could be grown abundantly enough to create a rich food supply. As we discovered how to grow a small number of crops, we made the natural world serve our purposes. Our species has become a turning point in the history of life on Earth.

Our lineage has always eaten figs, but it took six million years for us to learn how to plant a fig tree. This may seem like an oversimplified view of the origins of agriculture, but it underscores the idea that humans did not one day simply start farming. Agriculture was made possible by two things. First was the suite of physical, mental, social, and technological traits that had accumulated over the course of human evolution. Second was the rapidly thawing environment at the end of the last ice age, followed by a relatively stable climate.

Before the domestication of animals and plants, humans always lived by gathering, hunting, or scavenging. In the blink of an evolutionary eye, our species shifted to a dependency on the food we grow. In short order, we became a species that is fed, reliant mostly on food that is grown by a small percentage of the human population. *Homo sapiens* has been so successful in this new way of life that our numbers have exploded. As a result we are fundamentally altering Earth's ecosystems with consequences we are only now beginning to understand.

The height of the last ice age, which occurred between 27,000 and 19,000 years ago, marked the maximum spread of glacial ice sheets and the peak of the coldest, driest period of the past several hundred thousand years. After this peak, as the world became warmer and wetter, people encountered newly fertile, productive lands. Hunter-gatherers were already expert in the qualities of plants

Opposite: Ordering landscapes for human purposes, agriculture feeds billions of people, with consequences for biodiversity and water supplies.

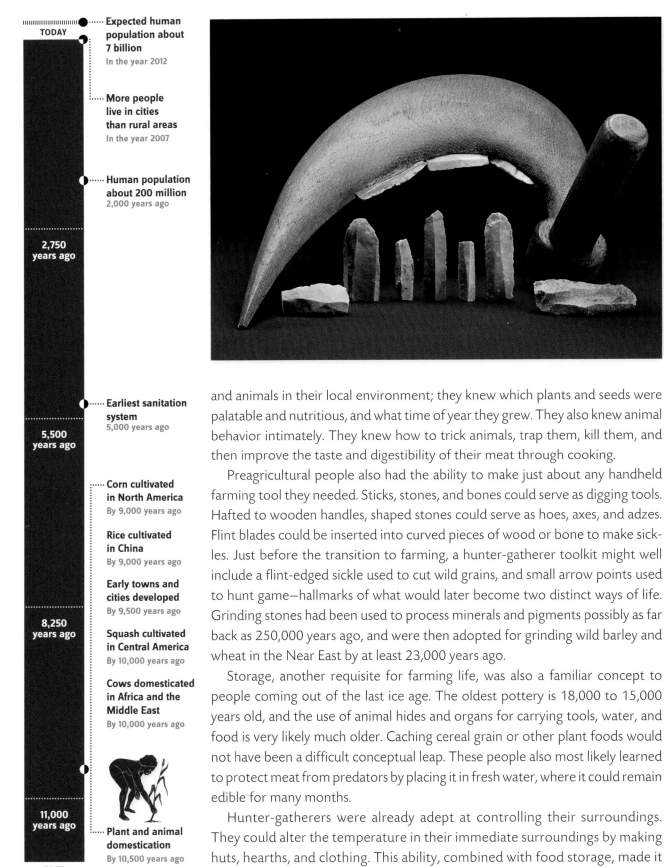

TODAY

Expected human population about 7 billion
In the year 2012

More people live in cities than rural areas
In the year 2007

Human population about 200 million
2,000 years ago

2,750 years ago

Earliest sanitation system
5,000 years ago

5,500 years ago

Corn cultivated in North America
By 9,000 years ago

Rice cultivated in China
By 9,000 years ago

Early towns and cities developed
By 9,500 years ago

8,250 years ago

Squash cultivated in Central America
By 10,000 years ago

Cows domesticated in Africa and the Middle East
By 10,000 years ago

11,000 years ago

Plant and animal domestication
By 10,500 years ago

PAST

and animals in their local environment; they knew which plants and seeds were palatable and nutritious, and what time of year they grew. They also knew animal behavior intimately. They knew how to trick animals, trap them, kill them, and then improve the taste and digestibility of their meat through cooking.

Preagricultural people also had the ability to make just about any handheld farming tool they needed. Sticks, stones, and bones could serve as digging tools. Hafted to wooden handles, shaped stones could serve as hoes, axes, and adzes. Flint blades could be inserted into curved pieces of wood or bone to make sickles. Just before the transition to farming, a hunter-gatherer toolkit might well include a flint-edged sickle used to cut wild grains, and small arrow points used to hunt game—hallmarks of what would later become two distinct ways of life. Grinding stones had been used to process minerals and pigments possibly as far back as 250,000 years ago, and were then adopted for grinding wild barley and wheat in the Near East by at least 23,000 years ago.

Storage, another requisite for farming life, was also a familiar concept to people coming out of the last ice age. The oldest pottery is 18,000 to 15,000 years old, and the use of animal hides and organs for carrying tools, water, and food is very likely much older. Caching cereal grain or other plant foods would not have been a difficult conceptual leap. These people also most likely learned to protect meat from predators by placing it in fresh water, where it could remain edible for many months.

Hunter-gatherers were already adept at controlling their surroundings. They could alter the temperature in their immediate surroundings by making huts, hearths, and clothing. This ability, combined with food storage, made it possible for groups to stay in one place year-round. Humans had also learned to burn landscapes to clear the underbrush, encourage new plant growth, or herd game.

Trade networks were already established for obsidian and other rock sources, along with exotic items such as seashells used to make beads. These earliest trade routes set the stage for the later spread of domesticated wheat and cattle as well as for knowledge about how to cultivate fields, clear pastures, and breed animals.

By the end of the last ice age, group cohesion, a prerequisite for permanent settlement, was already established through language. Symbolic behaviors such as tribal markings, ritual, and attire, which reinforced group identity and communicated individual status, were also in place. Those Paleolithic minds engendered something else indispensable for farming: the ability to imagine the future and plan ahead. With all these qualities in place, the transition to agriculture was feasible, if not inevitable.

THE TRANSITION

Archeologists long envisioned the transition from hunting and gathering to agriculture as a rapid process that occurred in a predictable step-by-step manner. It started with farming and then proceeded to herding, followed by population increase, which led to permanent settlements. Grinding of grains and the manufacture of pottery were hallmarks of this "Neolithic revolution"—an abrupt transition in only a few thousand years, which led to the oldest complex societies characterized by elaborate rituals and privileged ruling classes.

The latest archeological research, however, sees this transition differently. Rather than a rapid revolution, the process leading to agriculture and complex societies in the Near East, where it is best documented, was spread out from about 24,000 to 8,400 years ago. The interaction of post–ice age humans with their surroundings led to a mutual dependence between plants and humans, and ultimately domestication.

At first, hunter-gatherers could stabilize their food supply by combining their usual foraging activities with some cultivation of plants. As people tended more plants, they made a greater investment in particular plots of land, which made the mobile life of a hunter-gatherer less feasible. A dependence on domesticated foods probably started when weedlike cereals colonized disturbed areas around human habitat. Plants that were more palatable and had easy-to-use seeds or fruits would have been prime targets for human use. Ultimately, some cereals and plants, such as wheat, millet, barley, and fig trees, evolved so they could not reproduce without human help. Domesticated figs, for instance, were seedless and could reproduce only if a human took a cutting and planted it, making a clone. Before its domestication by humans, wild wheat fell to the ground to reseed. Once domesticated wheat was bred to remain upright, self-reseeding was not required to propagate, since humans were doing the job. If humans did not actively propagate such trees and seeds, their food supply would dwindle.

Opposite: *A model of a stone sickle holds predynastic Egyptian blades and is shown with blades from Ali Kosh, Iran, dated 8,600 to 8,000 years old. These tools show how ice age hunting technology was adapted for farming.*

A hunter-gatherer toolkit might well include a flint-edged sickle used to cut wild grains, and small arrow points used to hunt game—hallmarks of what would later become two distinct ways of life.

As the relationship between humans and plants evolved over time, the incentives for humans were significant: increased yields, drought tolerance, disease resistance, easier harvests, and better nutrition. As people spent more time in one place, any depletion in the wild vegetation would have made it advantageous to store and protect food crops. By developing a stationary food supply, people invested even further in settling down, cultivating the land, and maximizing the yields of specific food plants and herd animals.

A similar relationship developed between humans and animals. The incentives for animal domestication were to produce food and valuable commodities, such as wool, and to acquire assistance with work, such as carrying or pulling loads, plowing, or harvesting. Not just any animal could be domesticated. The best candidates had a social structure governed by a dominant individual, a tolerance of pens, little inclination to flee, a flexible diet, reduced aggression, prolific reproduction, and a fast growth rate. These factors narrowed the field greatly, ruling out many animals in Africa and the Americas. All told, of the top 20 domesticated animals, none evolved in sub-Saharan Africa, and only 2, the donkey and cattle, were present during the transitions to herding and farming in northeastern Africa. It is as if Africa's role as the crucible of humanity made it the least likely place for any unaggressive, unwary, and tolerant animal to evolve, as these traits would have been fatal in the face of hungry early human hunters.

DOMESTICATION

Plants and animals were domesticated in many different times and places in the world. The fact that people in different places independently began to practice the controlled breeding of plants and animals, and thus to change their appearance, nutritional qualities, and reproductive properties, further points to the readiness of hunter-gatherers in various parts of the world to intensify their use of the surroundings. In most regions of the world, plants were domesticated before animals, but exceptions include northeastern Africa, where cattle were domesticated prior to farming.

The first animal currently known to have been domesticated was the dog. Descended from the wolf about 16,000 years ago, dogs could help hunters and keep predators away from campsites and settlements. Goats were among the first animals domesticated as sources of food. By 10,000 years ago, people were selectively culling younger males and favoring a prolonged life in females, a pattern typical of human management of herds. Cattle were domesticated around the same time in northeast Africa and the Near East.

The cereal grains and pulses that make up the traditional "founder crops of agriculture"—including wheat, barley, peas, and lentils—were first cultivated, domesticated, and then farmed by about 10,500 years ago in the Fertile Crescent of the Middle East and southwest Asia. There is no evidence for a single center of agricultural origin; agriculture arose in widely separated geographic and climatic areas.

Farming's mixed blessing

Agriculture has had a profound impact on humanity. Innovations that included domestication and herding permitted humans to harness increasing amounts of energy and to support unprecedented population growth. Our present way of life depends on the production of food. Yet, as in other aspects of the human story, change comes at a price. The costs include negative impacts on our health, other species, and the environment.

Once humans began to live in closer proximity to each other and to animals, it became easier for diseases to spread. A sedentary lifestyle with high carbohydrate consumption and highly processed foods is also associated with health problems. Farming has significantly altered landscapes all over the world, as well. The collective biomass of cattle, sheep, pigs, and other large farm animals we support also adds substantially to the human footprint of managed land, wastes, and environmental alteration.

Based on molecular genetic and archeological evidence, rice was also domesticated early on, between 10,000 and 9,000 years ago in East Asia. Older rice grains from archeological sites have been found in China and South Korea, dating between 15,000 and 12,000 years old, but these were most likely the wild precursors.

By 8,000 years ago, wheat farming had reached the banks of the Nile River, and extensive agricultural lands occurred independently in the Far East with rice as the primary crop. By about 7,000 years ago, the Sumerians of Mesopotamia developed large-scale, year-round agriculture focused on wheat, organized irrigation, and the use of a specialized workforce. In Sumer, food production had intensified and expanded on a grand scale, eventually leading to the first city-states and large-scale societies.

Where people grew wheat, rice, and corn, or herded animals, they were able to harness energy sources unlike ever before. Grasses, including cereal grains, are rich in cellulose, which makes these plants nearly impossible for humans to digest without extensive processing. By investing their time in technologies that enabled grains to be processed, people could harvest on a broad scale a widespread, rapidly growing food resource. They domesticated only a few grass species—those that were the most palatable, had the largest

Modern animal husbandry focuses on a few domesticated large animals, such as these pigs at a show in Tulsa, Oklahoma. In recent decades, the number of domesticated animals worldwide has risen dramatically.

seed sizes, and were easiest to process. Yet by favoring the spread of these plants over others, they made grasses edible and accessible on a large scale for the first time.

As for the grasses that could not be domesticated, we organized other animals to consume them for us, and then we ate those animals or used them in a variety of ways that increased the energy that people were able to extract from their surroundings. Settlement, domestication, and agriculture thus greatly

As humans settled together in villages, such as this 9,000-year-old settlement at Çatalhöyük in Turkey, they benefited from sharing food, labor, and social values, but soon encountered disease and malnutrition.

intensified our ability to exploit many kinds of grasses—the type of plant that had become the most abundant and widespread during the previous two million years.

The transition from hunting and gathering to animal and plant domestication followed by the full-time commitment to agriculture seems to have permanently altered the composition of the human diet. If access to meat from large animals, beginning around 2.6 million years ago, represented the first transformation in the hominin diet, the second has involved a reliance on cultivated starches, which has dramatically changed the foods people eat over most of the globe.

POPULATION AND DISEASE

Within a few thousand years of the development of agriculture, population density rose sharply in the permanent settlements. Urban areas became centers of innovation and cultural enrichment. Yet one consequence was the loss of mobility, which early humans had used for millions of years to fend off food scarcity. In some places, growing food and feeding a larger population demanded large-scale irrigation, which altered the distribution of water and the landscape in unexpected ways. During some droughts, irrigation was a lifesaver. During the worst times, the investment in fields, permanent settlements, and large waterworks left little flexibility for anything other than watching the crops fail and the herds die, leading to starvation. The transfer of water to croplands also left behind a heavy burden of mineral salts, precipitated from the water, which in some areas transformed verdant fields into useless terrain.

The successes of agriculture were also favorable to the spread of disease. Before farming, people who hunted, gathered, and fished lived in small, dispersed groups, so that the transmission of disease was very limited. The rapid growth of populations eliminated this natural protection. As people became committed to living in one place in greater numbers, it was no longer possible to leave refuse behind as mobile foragers do. Farmers and settlers became surrounded by growing amounts of waste, and livestock were carriers of disease: Cattle carry forms of tuberculosis, smallpox, and measles; pigs carry influenza viruses. The microbes causing these diseases evolved in a way that maximized their spread from one human to another and across groups. The first known system of sanitation drains and public baths was built around 5,000 years ago in urban areas of the Indus Valley of modern-day Pakistan and India. However, the crowding of people, livestock, and waste in many parts of the world led to our species' continuing evolutionary "arms race" with microbes.

Eventually, improved roads and ships and the increasing numbers of travelers—soldiers, merchants, and migrants, among others—meant that when disease struck one area, it was usually carried to others, which led to epidemics and set the stage for later worldwide pandemics. Between the years 1348 and 1351, for example, one-third of the European population—some 25 million people—died of bubonic plague; and in just one year between 1918 and 1919, 20 million to 40 million people died of flu.

THE TRANSFORMATION OF ECOSYSTEMS

Despite these negative consequences, the agricultural way of life expanded across the world and set the stage for our way of life today. As people crowded into cities, agricultural activities were called upon to provide a massive amount of food and other supplies. The transfer of energy from field to city was built upon the ancient human impulse to convey food from one place to another to feed others; but its modern expression was unlike anything foreseeable in the Stone Age.

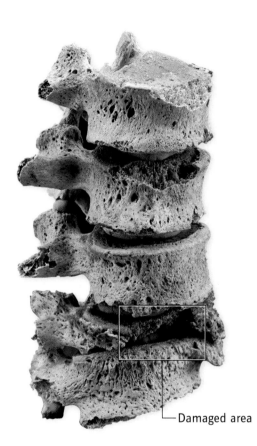

Damaged area

A section of vertebrae from a 3,200-year-old burial in Jordan shows damage from disease, probably tuberculosis in this case.

As large, complex societies came to rely on only one or a few types of crops like wheat, rice, or corn, people began to simplify the environment. This simplification is the origin of human-dominated ecosystems, characterized by a reduction in species diversity, with the intent of controlling landscapes for human purposes. The rise of human-dominated ecosystems signifies a profound reshaping of habitats.

The rapid impact of human-dominated ecosystems can be judged by looking far back in the history of life. At least 250 million years ago, when plant-eating vertebrates had their first major evolutionary expansion, land ecosystems began to exhibit a standard transfer of energy up the food chain: The grand diversity of plants was consumed by a smaller diversity of plant-eating animals, which were eaten by a still smaller number of carnivores.

Human-dominated ecosystems have dramatically altered this classic, pyramid-shaped food chain. The greatest flow of energy and the strongest controls on the system do not move between a vast diversity of plant and animal species; rather they move in relation to a single species, *Homo sapiens*. Plant and animal diversity has been sharply reduced as humans focus on a very small number of foods, and supplant forests, wetlands, and other habitats with fields, pastures, and places to put wastes and pollutants. The long-established top of the food chain has also nearly been eliminated as large carnivores are killed off, favoring only the tiniest of predators—the microbes that have proliferated along with humans, crop plants, and livestock that form the foundation of the new-style ecosystems.

Of all the ways of describing a human being—a bipedal primate, a large-brained toolmaker, a member of a symbolic species—the one most relevant to the dilemmas of the present is that we are a turning point in the history of life. At least 38 percent of the global land surface, excluding the polar ice caps, is devoted to farming, and only 17 percent of the world's land area has escaped direct influence by humans. It is estimated that the world population 2,000 years ago was about 200 million; 1,000 years ago, our numbers exceeded 10 million. In just the past 40 years, the human species has more than doubled to nearly seven billion people. In 2007, more people lived in cities than in rural areas for the first time, and urban dwellers are responsible for about three-quarters of the world's energy consumption. From 1961 to 2004, the population of cattle, pigs, sheep, and goats increased from 2.7 billion to 4.1 billion, and domesticated fowl from

> In 2007, more people lived in cities than in rural areas for the first time, and urban dwellers are responsible for about three-quarters of the world's energy consumption.

FAQ:
How do we know about the health of the first farmers?

Human skeletal remains suggest that the health of the first farmers differed considerably from the health of hunter-gatherers. On average, the first agriculturalists lived shorter lives and were smaller in stature. They also suffered from new nutritional deficiencies and infectious diseases, some of which left diagnostic marks on the skeleton in the form of lesions and other abnormalities. A growing emphasis on foods with higher carbohydrate content, like domesticated cereals, made early farmers' teeth more vulnerable to cavities. Their smaller jaws created dental crowding and problems associated with impacted teeth. Although agricultural life increased female fertility, the added nutritional drain of more pregnancies also had a negative effect on female health.

3 billion to 16 billion. The vast areas of land needed for these animals and the foods they require directly compete with the basic necessities of wild species.

From one perspective, this fundamental transformation of Earth's ecosystems is the culmination of the ability of humans to adapt in the face of an unstable world. By our gaining control over the food supply, we have expanded enormously the footprint of human activity. To make sense of this process, though, it is important to recall where we started this chapter: The transition to an agricultural way of life began as human populations came out of the last ice age. After a brief return to cold conditions between 12,800 and 11,600 years ago, world climate began to stabilize. Over the past 10,000 to 8,000 years, Earth's climate system has been unusually calm, with only minor fluctuations documented in climate records from the deep-sea and Greenland ice cores. As noted in previous chapters, evolution over the long haul instilled in humans a suite of adaptations that enabled our lineage to adjust or recover from climatic perturbations; yet over the past several thousand years, these capacities have been expressed in a peculiar time of remarkable stability.

The pace at which humanity grew in numbers and advanced culturally and technologically over the past 10,000 years is perhaps matched only by the rate at which we have altered—some would say destroyed and squandered—our surroundings. Yet if there is one thing we learn from the longer-term perspective, it is that the environment eventually changes, and we cannot always control what occurs. That said, humans now seem to have become as close to masters of our fate as any creature ever was on Earth. The question is: Can we handle it?

Soaring forest canopy dwarfs a person in Costa Rica. Our knowledge of Earth's biodiversity grows alongside our awareness of how rapidly we are losing it.

ARE WE IT?

W HAT SIGNIFICANCE SHOULD WE PLACE ON THE DEMISE OF EARLIER HUMAN species? Some had large brains; all were highly social. Some made tools with great care and possessed the fundamentals of symbolic thought. Each lineage developed, culled, and compounded certain essential qualities of what it has meant to be human. If we could understand the persistence of some of these species and the extinguishing of others, we might learn something valuable about the origin and conceivable future of our own.

The question "Are we it?" appears to invite conjecture as to whether humans are the end goal or pinnacle of evolution. Yet that approach to the question reflects an outmoded view of human origins. Our evolutionary history has been an intricate dance between the qualities that aided the survival of our ancestors and the ever changing surroundings that relentlessly tested the effectiveness of those adaptations. Over the course of human evolution, many species have come and gone. While some were the direct ancestors of our own kind, the human family tree is replete with species and ways of life that no longer exist. Bearing this in mind, a different way to interpret the title question is to ask whether we are the last of the hominins, the end of the bipedal branch of the primate family tree. "Are we it?" invites comment as to how far into the future *Homo sapiens* will survive. And if we do survive, can we thrive?

There is a pattern that can be detected in studying those close evolutionary cousins of ours who are known now only through the fossil evidence. In the face of environmental change, extinction appears to have been the fate of species that followed a narrow course of adaptation to their surroundings. They may have been well suited for living in a limited range of environments, or they may have found themselves confined to a small geographic area with little margin for surviving difficult times. Other species making up our family tree took advantage of a wider geographic range or ate a variety of foods but ultimately were saddled with certain qualities that thwarted their resilience when new conditions arose.

The last known Neanderthal populations had physical adaptations and behaviors that enabled them to exploit the cool, wooded habitats of western Europe. Beginning around 33,000 years ago, the Neanderthals

Opposite: *Humanity stakes its claim to every patch of habitable land in Rio de Janeiro, Brazil, whose metropolitan area is home to more than ten million people.*

Earth at night glows brightly in urbanized areas but shades toward darkness where humans are fewer or electric power is scarce, in an image compiled from satellite and ground data.

remained south of the last major expansion of glaciers and tracked the migration of cool woodlands to Iberia and Gibraltar. There they flourished one last time, until possibly as late as 28,000 years ago. During that era, groups of *Homo sapiens* arriving from tropical areas proved capable, through cultural adaptations, of surviving in colder climates than even the Neanderthals could stand. Modern humans thus gained a foothold they would never relinquish in the zones where Neanderthals would have expanded, if they could have, when warmer times returned. It's not that Neanderthals weren't well adapted to cold conditions; of all the hominin species, they had the bodies best equipped to deal with the cold. Yet the adaptable, resilient *Homo sapiens* could flourish in the frigid zones of ice age Europe and also in the cool, the warm, and the hot climates they

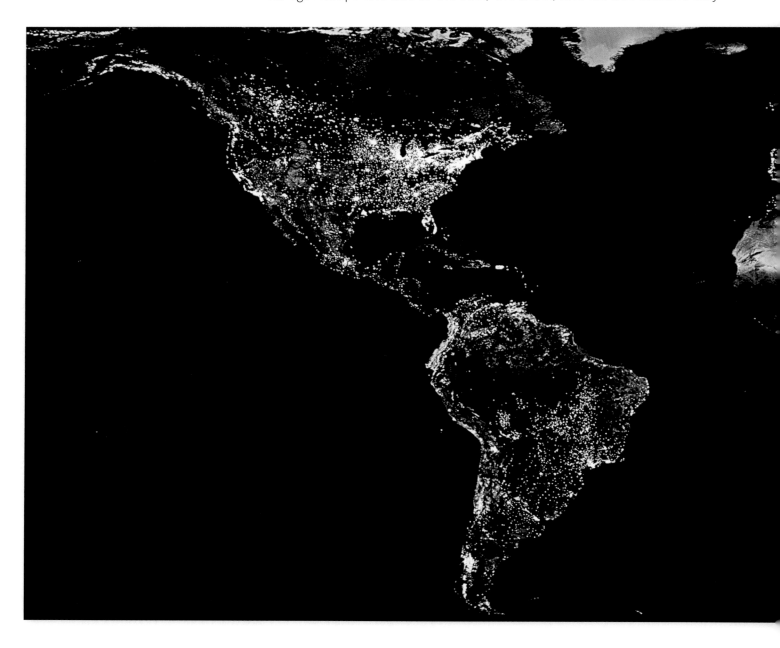

encountered elsewhere. *Homo neanderthalensis* was an example of a species at its best within a limited range of environments.

An example of a species that was confined geographically is *Homo floresiensis,* which persisted for possibly hundreds of thousands of years in Indonesia. As far as we know, the only adaptive options open to the "hobbit" were those limited possibilities available on a small island. There, on the island of Flores, it went extinct by about 17,000 years ago. As is true for species today and in the recent past, the smaller the numbers and the smaller the inhabited area, the more likely extinction is to occur.

A similar fate befell *Homo erectus,* which thrived over a vast area for more than 1.5 million years. Its successful run ended after its range narrowed to the island of Java, where fossils of the last of its kind have been found, dated as late as 70,000 years ago.

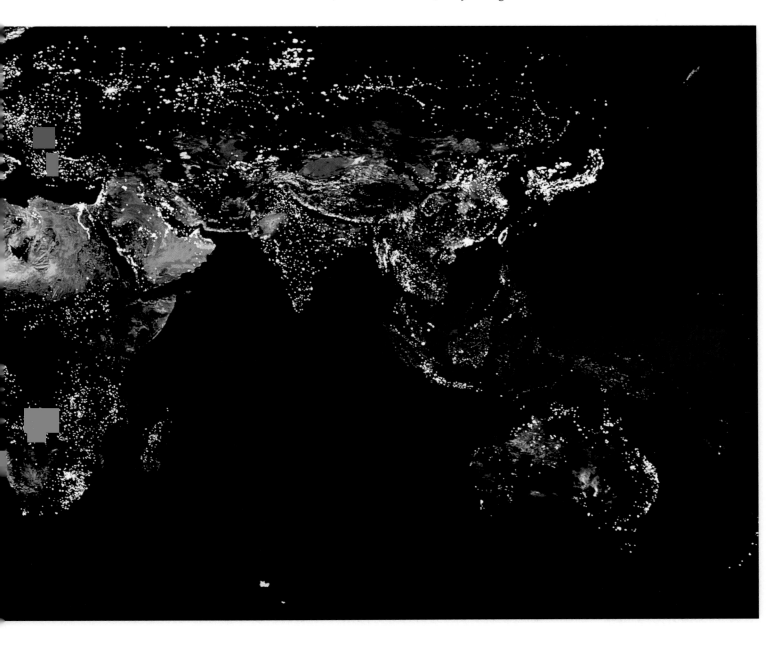

Opposite: *Malaria destroys red blood cells. Humans are striving to develop the technology to rid the world of malaria and many other scourges.*

The human effects on climate via CO_2

Humans do not simply adapt to Earth's climate, we know now. We can also change it, through the burning of fossil fuels and other activities that contribute to a rapid rise in atmospheric levels of greenhouse gases, particularly carbon dioxide (CO_2). Research shows that the concentration of atmospheric CO_2 correlates with average global temperature. Atmospheric CO_2 has risen from 280 to 385 parts per million (ppm) since preindustrial 1750. This is far above Earth's normal range of 180 to 300 ppm over the past 650,000 years. Global temperature rose about 0.74°C (1.33°F) over the past century. An additional 0.4°C (0.72°F) temperature increase is expected over the next two decades.

Predicted increases in CO_2 and other greenhouse gases are expected to have many severe consequences. One of the most compelling questions of our time is how 21st-century humans will respond to rising sea levels, increased intensity of storms and droughts, and other threats associated with global warming.

A final example of the overall pattern is *Paranthropus boisei,* which thrived in eastern Africa for at least one million years. The reasons for its demise are hard to fathom, yet we may note the hallmark of this species—a chewing apparatus so impressive that it could eat almost anything it wished. Several scientific studies indicate that its diet was quite broad, and it could stave off starvation by eating tough, coarse foods. Yet this so-called "nutcracker species" may also have been encumbered by its own adaptation. Even the softest berries or insects required the mobilization of its heavy-duty chewing equipment. After such a successful run, the fall of *P. boisei* brings up the possibility that its powerful jaws, massive muscles, and broad teeth ultimately posed more of a constraint than a license to adjust to new kinds of food.

The broad theme here is the importance of a certain degree of flexibility if organisms are to respond successfully to environmental ups and downs. This, in turn, raises questions about aspects of an organism's way of life that might actually encumber it under new circumstances. A species that does a lot of things well retains more options when its surroundings change. This general rule applied to our bipedal kin, and there is every reason to suspect it applies to *Homo sapiens* as well. What developments in the world today serve to promote, or limit, our own species' adaptability? What is the nature of our flexibility—and our constraints?

UNINTENDED EXPERIMENTS

A starting point is to try to understand the challenges we create for ourselves as a result of our species' profound influence on the world today.

In tracking the course of human evolution, we see that the global spread of *Homo sapiens* resulted from our evolving capacity to alter our immediate surroundings. By chipping the edges of rocks, *Homo habilis* began to modify its surroundings. The ability to control fire, make a hearth, and build a shelter meant that *Homo heidelbergensis* could cook its food and alter the temperature of its primitive living quarters. Later still, people used fire to alter entire landscapes; fire could move animal herds and help regenerate vegetation. The evolution of symbolic communication also heightened our species' ability to manipulate our surroundings. Symbols and language made it possible for social networks to grow and for resources and information to be exchanged among groups, providing buffers in lean times and pathways for innovation. Language gave us the tools to plan large-scale activities, react to change, and imagine things beyond our reach.

The cumulative effect of these small steps was a way of life so successful that it enabled people to spread around the globe. The simple idea of nurturing an edible plant or keeping docile animals together in one place helped secure the local food supply and led to sensible decisions to domesticate certain plants and animals and promote their use at the expense of others. These local successes set the stage for environmental change everywhere—an unintended but real consequence of a long and complex evolutionary history. Considering the

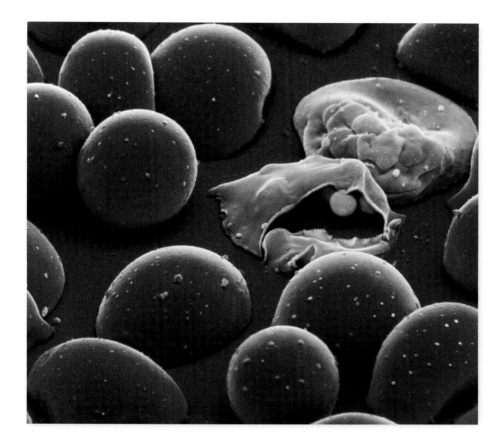

We can see that the stresses our activities now place on landscapes and other species could constrain our future.

six million years of our evolutionary past, all of these later developments took place very quickly.

Virtually all species alter their immediate surroundings in some manner; yet how might the power and breadth of human influence affect our own resilience? As population has ballooned with the rise of modern societies, we can see that the stresses our activities now place on landscapes and other species could constrain our future. In many parts of the world, soil nutrients and fresh water have become degraded, even though our way of life depends on them. According to the 2005 report *Millennium Ecosystem Assessment,* out of more than 10,000 edible plants, fewer than 20 species produce most of our food, and only three crops—wheat, rice, and corn—provide half the calories consumed worldwide. Similar statistics apply to animals: Of an estimated 15,000 species of mammals and birds, fewer than 14 account for 90 percent of human consumption of animal products.

This is specialization, a narrowing of possibilities. It means that economic decisions ignore the broad foundation of plant and animal species that underlies our food options and the functioning of ecosystems in which all organisms live. Being ignored, unfortunately, means that the necessities on which most species depend are threatened or have already been eliminated. Fields, fences, and urban zones diminish the ability of most organisms to move in response to environmental change, thus limiting the long-evolved adaptability of many species. The impact of *Homo sapiens* also reaches the ocean, where populations of fish, shellfish,

Global warming is evident in receding glaciers and melting ice sheets, such as this one in Greenland. How we respond to change, including changes we may be causing, will be a determining factor in our future.

and other species have been lost or depleted. According to recent assessments, the current instability of marine ecosystems because of human use will have a profound effect on ocean food production, shrinking this rich source of biological diversity and our ability to use resources from the sea.

A further challenge comes from the effort to support an enormous and growing human population. Our obligation to feed ourselves and others magnifies the need to transform landscapes for our own purposes on a global scale. Our reliance on food production initially proved beneficial because it could support small numbers of people in villages and towns. But now it must provide for massive urban areas where consumers far outnumber those who actively produce food. It is, so far, a short experiment, but we do know that about 83 percent of the Earth's viable land surface has already been influenced directly—converted to agricultural fields, urban zones, waste storage depots, and human-controlled landscapes altered by deforestation, mining, and pollution. By 2007, humans had also built so many dams that the amount of water held in storage was nearly six times greater than in free-flowing rivers. Our species is so firmly committed to these ways of altering our surroundings that it is difficult to see what room we have to maneuver and what options will be available in the future.

The way in which greenhouse gases released by the burning of fossil fuels affect the atmosphere, and thus climate, is now well documented. Over time, atmospheric fluctuations in carbon dioxide, the most significant greenhouse gas, have closely tracked oscillations in Earth's temperature and the expansion and melting of the ice caps. But human activities have now pushed the atmospheric concentration of carbon dioxide to about one-third higher than it has been at any time since the origin of *Homo sapiens.*

According to even moderate estimates, the concentration of carbon dioxide in the atmosphere over the next century is projected to be twice as high as it has been at any time since the origin of the human lineage six million years ago. This rise implies a considerable temperature increase and at least a half-meter elevation in sea level, which would inundate low-lying coastal regions where one-tenth of the world's population now lives. Our species will experience environments and rates of change never previously encountered in our evolutionary history.

Evolution and the history of living things are at the heart of understanding the resilience of Earth's environments and its species, and our own sources of adaptability.

Earth's climate system rarely responds like a barometer, with a clear trend one way or another that allows us to predict exactly what will occur. Instead, climatic responses occur as thresholds; by the time the indications become obvious, the change has essentially already occurred. The forewarnings of climate warming are serious, yet the challenges ahead are possibly even more daunting than those posed by any definable trend or predictable problem, no matter how startling. We are in the midst of a giant experiment as our species creates new and inadvertent inputs that brush up against Earth's climate instability. A volatile climate and the insecurity it causes may thus present a stronger test than any trend we may anticipate.

SOURCES OF OUR RESILIENCE

There is no easy answer to how our species will live or whether we will thrive in the future. The guideposts are few, yet they include the processes of survival, change, and extinction that shaped the fate of species most like us—those versions of "human" known from the fossil record.

A fundamental concern for modern society is whether the core adaptations of human beings can help us solve and adjust to the trials of environmental

---

FAQ:
Are humans still evolving?

There is strong evidence that human populations are still evolving. Since the innovation of animal husbandry, for example, some groups have inherited a mutation that allows them to digest milk long after infancy. This trait evolved independently in two areas' populations —6,000 to 5,000 years ago in Europe, and 3,000 years ago in Africa.

Another piece of evidence that humans are still evolving came in the 1940s, when doctors discovered that patients with sickle-cell anemia were more likely to survive malaria. Although two copies of the sickle-cell gene (inherited from both parents) cause sickle-cell disease, one copy (inherited from one parent only) confers malaria resistance.

168 | WHAT DOES IT MEAN TO BE HUMAN?

Opposite: *A transplanted eucalyptus seedling signals hope for reforesting a clear-cut tract in Indonesia—and the importance of our own actions for humanity's future.*

change that we now face. As we wrestle with this question, a lot can be learned by investigating the workings of the natural world and human interactions with it. Evolution and the history of living things are at the heart of understanding the resilience of Earth's environments and its species, and our own sources of adaptability.

What we have learned here is that the defining characteristics of our species evolved as Earth's environment shifted again and again, the fluctuations widening over time. This perspective on human origins suggests that the lineage that made it to the present has inherited a repertoire of mental and social tools that can help it adapt to change and uncertainty. In the intricate social worlds we inhabit, the human brain compels us to evaluate probabilities and potentials, to create a universe of opinions and actions toward others. The creativity of language imparts in us the ability to respond swiftly to unforeseen events and their consequences. The capacity of humans to innovate lays the groundwork for building new technologies and opportunities. Our beliefs and ethical capabilities will come into play as the human species adapts to the future through an evolutionary process that we, to a certain degree, can influence and direct. The creative capacity to diversify culturally has led to the multitude of ways in which human beings live. Cultural diversity and the inclination to adopt differing viewpoints within a society have multiplied many times over the possibilities and options of our species.

The evolution of human adaptations during one of the most volatile eras in Earth's climate history means that agility in responding to changing conditions is part of our makeup. The biological sciences used to frame human nature as a channeling of human possibilities within genetic constraints. *Inevitability* was the keynote of that view of nature. A different perspective is taking shape, however, in which certain unique qualities of human life—the activity of the brain, the unavoidable learning of language, an appetite for innovation, an aptitude for abstract thought—create a certain resourcefulness, a nimble capacity to adjust to the conditions at hand. *Adaptability* is the keynote of this view of human nature.

Scientific exploration of human origins shows us that the human species is adapted to change. When conditions are altered, we employ our mental flexibility and our social tendencies to respond—in some instances, by calling upon our deep capacity to assist others, and at other times, by acting violently toward those we perceive as threatening or different. The balance that we strike between these tendencies may decide much about the prospects ahead. Our evolved characteristics offer the chance to imagine the future, to adjust our thinking, to bring to bear our unique potential for caring and our need for meaning. These qualities emerged during the long common ancestry shared by all people, and may be a source of hope in meeting the challenges and adversity of the future. Perhaps these qualities will ultimately define our species' answer to what it means to be human.

GLOSSARY

Adaptation: A feature produced by natural selection for its current function

Australopith: Member of a species in the genus *Australopithecus*

Bipedal: Habitually walking upright on two legs

Brow ridge: Bony ridge above the eye sockets

Calcareous: Composed of calcium carbonate

Cerebral cortex: Outermost layer of the cerebrum, typically consisting of "gray matter"

Cerebrum: Largest part of the brain, including the frontal, occipital, temporal, and parietal lobes

Chromosome: Organized package of DNA and proteins found in a cell nucleus

Core: Source stone reduced in size by intentional removal of flakes

Cortex: Equivalent to the cerebral cortex

Cytoplasm: Thick, jelly-like substance that fills a cell

Domestication: Human-induced artificial selection, taming, or breeding of plants or animals

Environmental variability hypothesis: The hypothesis that adaptation to a variable environment, rather than a static environment or directional change, has characterized human evolution

Evolution: Descent with modification

Flake: Sharp piece of stone intentionally removed from a core

Foraminifera: Single-celled microorganisms with calcareous shells

Frontal lobe: Most forward part of the cerebrum

Gene: The unit of heredity; a region of DNA with a particular observable effect

Genera: Plural of genus, the rank above species in Linnaean classification

Genome: All the genetic information in an organism

Greenhouse gases: Atmospheric gases that trap heat, warming Earth's surface

Hammerstone: Cobble used to strike flakes from a stone core

Hominin: The human evolutionary group of species, both fossil and modern

Hypothesis: A proposed, testable scientific explanation for a particular set of phenomena

Lithic: Consisting of stone or rock

Lumbar: Relating to the lower back and its vertebrae

Mandible: Lower jaw

Metabolism: Chemical reactions necessary to maintain life

Microcephaly: Neurodevelopmental disorder resulting in an abnormally small head and brain

Miocene: Geologic time period ranging from about 23 million to 5.3 million years ago

Mitochondria: Parts of a cell that generate most of its chemical energy

Molecular: Having to do with DNA sequences or the amino acid sequences of proteins

Mutation: Change in a DNA sequence

Natural selection: Differential survival or reproduction in a population leading to change in its genetic makeup

Neocortex: Outer layer of the cerebral hemispheres of the brain

Neolithic: Last part of the Stone Age, before the origin of metal tools

Neural: Relating to a nerve or the nervous system

Nucleus: Part of a cell containing genetic material

Pliocene: Geologic time period ranging from about 5.3 million to 1.8 million years ago—or, for some researchers, to 2.6 million years ago

Primates: The biological order of mammals containing lemurs, lorises, galagos, tarsiers, monkeys, and apes

Species (singular and plural): All members of a population or set of populations that actually or potentially interbreed over time

For other definitions and information about evolution, see
http://evolution.berkeley.edu/evolibrary/glossary/glossary.php

AUTHOR ACKNOWLEDGMENTS

RICHARD POTTS: I am most grateful for the dedicated efforts and support of many Smithsonian colleagues, particularly Jennifer Clark, Briana Pobiner, Matt Tocheri, Kay Behrensmeyer, Alison Brooks, Brian Richmond, Bernard Wood, Elizabeth Jones, and Chip Clark; and the creative talents of our exhibition core team, including Sharon Barry, Junko Chinen, Kathleen Gordon, Myles Gordon, and Michael Mason. I wish to recognize the extraordinary artwork of John Gurche and Karen Carr. I also thank Chris Sloan and the National Geographic Society staff for their diligent efforts on this book. I give special thanks to colleagues and institutions around the world for their contributions to the exhibition.

CHRISTOPHER SLOAN: I am grateful to scientists, like Rick Potts, who have generously shared their thoughts with me over the years. I am also grateful to National Geographic Society colleagues such as photographer Kenneth Garrett, writers Rick Gore and Jamie Shreeve, and editor Hannah Bloch for hours spent discussing human evolution and reviewing my drafts for this book. The NGS book division's Sanaa Akkach, Adrian Coakley, Lisa Thomas, and Susan Hitchcock made the book a great read and visual treat, as did editor Leslie Allen. And thanks to Elizabeth Jones for her great project management. Finally, I thank my family, who have put up with my many late nights and working weekends.

The following Internet address credits the numerous partner organizations and individuals who contributed to the David H. Koch Hall of Human Origins and the Human Origins Initiative of the Smithsonian Institution's National Museum of Natural History:
http://humanorigins.si.edu/about/acknowledgments

ILLUSTRATION CREDITS

Notes: * Denotes specimens that are casts or scanned 3D replicas. For more information about illustrator Karen Carr, see www.karencarr.com. For information about specific photographs by Chip Clark, Donald E. Hurlbert, and James F. Di Loreto (staff at the National Museum of Natural History, Smithsonian Institution) please reference the photo number listed in the credits. Cover, Don Hurlbert & Jim DiLoreto (2009-27196); 2-3, Michael Nichols; 4, David L. Brill; 7, Chip Clark; 9, Erich Lessing/Art Resource, NY; 10, William Perlman/Star Ledger/CORBIS; 11, mouse: David Tipling/Getty Images; chimpanzee: Michael Nichols; gorilla: Jason Edwards/NationalGeographicStock.com; orangutan: Konrad Wothe/Minden Pictures/NationalGeographicStock.com; rhesus monkey: Richard T. Nowitz/NationalGeographicStock.com; humans: Allen Russell/Photolibrary; Con Tanasiuk/Photolibrary; Baymler/Getty Images; Elizabeth Young/Getty Images; Dougal Waters/Getty Images; Asia Images Group/Getty Images; bananas: Chris Windsor/Getty Images; chickens: MIXA/Getty Images; 13, large torso: Robert Clark; small torso: CORBIS; ear: Paul Sutherland/NationalGeographicStock.com; finger: Joel Sartore/NationalGeographicStock.com; knees: Gallo Images/Getty Images; head: Becky Hale/NationalGeographicStock.com; 15, Ken Eward; 17, icons: Karen Carr & David Hsu; 18-19, Michael Nichols; 21, Cyril Ruoso/Minden Pictures/NationalGeographicStock.com; 22, lemur: Nicole Duplaix/NationalGeographicStock.com; tarsier: Tim Laman; new world monkey: Roy Toft/NationalGeographicStock.com; old world monkey: Bates Littlehales; lesser ape: Cyril Ruoso/Minden PIctures/NationalGeographicStock.com; orangutan: Konrad Wothe/NationalGeographicStock.com; gorilla: Jason Edwards/NationalGeographicStock.com; chimpanzee: Michael Nichols; humans: Allen Russell/Photolibrary; Con Tanasiuk/Photolibrary; Baymler/Getty Images; Elizabeth Young/Getty Images; Dougal Waters/Getty Images; Asia Images Group – Getty Images; icons: Karen Carr; 23, Frans Lanting; 25, Michael Poliza/NationalGeographicStock.com; 26, Karen Carr; 28, Chip Clark (* 2009-41354); 31, David McLain; 32-33, art: Karen Carr; humans: Allen Russell/Photolibrary; Con Tanasiuk/Photolibrary; Baymler/Getty Images; Elizabeth Young/Getty Images; Dougal Waters/Getty Images; Asia Images Group/Getty Images; 35, Enrico Ferorelli; 36-37, art: John Gurche, photos: Chip Clark; 38-39, Chip Clark (* Australopithecus africanus, STS5: 2009-41330); (* Homo rudolfensis, KNM-ER 1470: 2009-41267); (* Homo erectus, Sangiran 17: 2009-41308); (* Homo heidelbergensis, Petralona: 2009-41241); (* Homo sapiens, Fish Hoek: 2009-41348; 40, Kenneth Garrett; 43, Chip Clark; 45, Robert W. Madden; 46, Suzi Eszterhas/NationalGeographicStock.com; 47, Chip Clark; 48, Karen Carr; 49, Jim Brandenburg/Minden Pictures/NationalGeographicStock.com; 51, Richard Potts, Smithsonian Institution; 52-53, Chip Clark (* Homo neanderthalensis, Shanidar 1: 2009-41230 & 41363); (* Australopithecus africanus, Taung child: 2009-41403 & 41419); (* Homo heidelbergensis, Kabwe 1: 2009-41254 & 41429); 54, Karen Carr; 55, Chip Clark (* Papio hamadryas: 2009-41393); Donald E. Hurlbert & James F. Di Loreto (* Paranthropus robustus cranium & mandible: 2009-27200 & 2009-27204); (* tools: 2009-27220); (* Paranthropus robustus skull & Panthera pardus: 2009-27225); illustration: Jay Matternes; 56-57, Chris Johns; 59, Skip Brown/NationalGeographicStock.com; 60, Karen Carr; 61, Bob Campbell; 63, Chip Clark (* 2009-41823); 64, art: Karen Carr; photos: Chip Clark (Sahelanthropus model: 2009-41332); (* Australopithecus anamensis: 2009-41602); (* Homo erectus: 2009-41599); 66, © John Gurche 2009; 69, James A. Sugar; 70, Karen Carr; 71, Michael Nichols; 72-73, © John Gurche 1987; 74 (UP), Karen Carr; 74 (LO), Belinda Wright; 76, Zeresenay Alemseged/Fred Spoor/Courtesy National Museum of Ethiopia; 77 (LE), Karen Carr; 77 (RT), Chip Clark (2009-41533); 78, infant chimpanzee: Stan Osolinski/Photolibrary; adolescent chimpanzee: Dorling Kindersley/Getty Images; adult chimpanzee: Daryl Balfour/Getty Images; human baby: Hisayoshi Osawa/Getty Images; human child: Tyler Marshall/Getty Images; human teenager: Peter Augustin/Getty Images; human adult: Sheer Photo, Inc/Getty Images; human senior: Benjamin Rondel/Corbis; 79 (LE), Takeru Akazawa; 79 (RT), Chip Clark (* 2009-29122); 81, Wilbur E. Garrett; 82 (LE), icon: Karen Carr; 82 (RT), Chip Clark (2009-41774)—top: courtesy of Dr. Jill Pruetz; middle: courtesy of Prof. Tetsuro Matsuzawa; bottom: courtesy of Dr. Kathelijne Koops and Dr. William McGrew; 83, Photo courtesy of Professor Tetsuro Matsuzawa, Kyoto University Primate Research Institute, Kyoto, Japan; 84, (LE) Chip Clark (hammerstone: 2009-41703); (* core: 2009-41741); (* flake: 2009-41742); 84, (RT) Karen Carr; 86, Richard Potts; 87, David L. Brill; 88, Jason Nichols; 91, Sebastien Starr/Getty Images; 92 (LE), icon, Karen Carr; 92 (RT), © John Gurche 2009; 94, Chip Clark (* Australopithecus afarensis, AL 2881: 2009-41819); (* Homo erectus, KNM-WT 15000: 2009-41642); (* Homo neanderthalensis: 2009-41615); 96, Karen Carr; 97, Gordon Wiltsie/NationalGeographicStock.com; 98, Kenneth Garrett; 99, Ira Block; 101, Cary Sol Wolinsky; 102, Karen Carr; 103, Ira Block; 105, Donald E. Hurlbert & James F. Di Loreto (* Australopithecus afarensis: 2009-27252); (* Homo rudolfensis: 2009-27221); (* Homo erectus: 2009-27222); (* Homo heidelbergensis: 2009-27258); (Homo sapiens: 2009-27223); 107, Kenneth Garrett; 108, Jason Edwards/NationalGeographicStock.com; 111, Jon T. Schneeberger and Larry Kinney; 112, Karen Carr; 113, Chip Clark (* Elephas recki mandible: 2009-41362); (* Equus oldowayensis: 2009-29952); (* Homo erectus: 2009-29933); (tools: 2009-29928); (* Elephas recki rib: 2009-29914); 114-115, Jodi Cobb/NationalGeographicStock.com; 117, Stephanie Maze; 118 (LE), icon, Karen Carr; 118 (RT), KEENPRESS; 119, Kenneth Garrett; 120, Chip Clark (left to right: 2009-41711; 2009-41712; 2009-41716; 2009-41719; 2009-41718); 122-123, Chip Clark (2009-41773)—(left to right: * needles—5; perforator; * barbed points—2; * harpoon; burins—3; * engraved bone; * spear thrower; * stone points—bottom 2; * stone points—top 2: Blombos Cave stone tool replicas, courtesy of Iziko Museums, Cape Town, Republic of South Africa); 124, Karen Carr; 125, Chip Clark (2009-41497); 127, George Steinmetz; 128 (LE), icon: Karen Carr; 128 (RT), Chip Clark (* 2009-29062)—(engraved ocher plaque replica, Blombos Cave, courtesy of Iziko Museums, Cape Town, Republic of South Africa); 129, Thomas J. Abercrombie; 130 (LE), Chip Clark (Palette: 2009-41687); (Hematite: 2009-10124—Courtesy of The Stone Age Institute); 130 (RT), Art Resource, NY; 131, Chip Clark (2009-41665); 132, Karen Carr; 134-135, KEENPRESS/NationalGeographicStock.com; 137, Kenneth Garrett; 139, Mike Theiss//NationalGeographicStock.com; 140, icon, Karen Carr; 141, Donald E. Hurlbert & James F. Di Loreto (* Homo sapiens, Irhoud 1: 2009-27254); (* Homo sapiens, Liujiang: 2009-27237); (* Homo sapiens, Kow Swamp: 2009-27227); 142, NG Maps; 144, David Liittschwager; 145, art: John Gurche, photos: Chip Clark; 146-147, Sarah Leen; 148, Karen Carr; 149, Chip Clark (scraper: 2009-41376); (soil: 2009-41375); (* Homo neanderthalensis, Shanidar 4 mandible: 2009-41529); (* Homo neanderthalensis, Shanidar 4 hand: 2009-41675); 150, Jim Richardson; 152 (LE), icon, Karen Carr; 152 (RT), Chip Clark (2009-41691)—(* sickle; blades—10); 155, Joel Sartore; 156, Manoocher Deghati; 157, Chip Clark (2009-41494); 159, Michael Nichols; 161, Stephanie Maze; 162-163, Earth Imaging/Getty Images; 165, Albert Bonniers Forlag AB; 166, James D. Balog; 169, James P. Blair.

INDEX

WHAT DOES IT MEAN TO BE HUMAN?
RICHARD POTTS AND CHRISTOPHER SLOAN

PUBLISHED BY THE NATIONAL GEOGRAPHIC SOCIETY

JOHN M. FAHEY, JR., *President and Chief Executive Officer*

GILBERT M. GROSVENOR, *Chairman of the Board*

TIM T. KELLY, *President, Global Media Group*

JOHN Q. GRIFFIN, *Executive Vice President;
    President, Publishing*

NINA D. HOFFMAN, *Executive Vice President;
    President, Book Publishing Group*

PREPARED BY THE BOOK DIVISION

BARBARA BROWNELL GROGAN, *Vice President and
    Editor in Chief*

MARIANNE R. KOSZORUS, *Director of Design*

CARL MEHLER, *Director of Maps*

R. GARY COLBERT, *Production Director*

JENNIFER A. THORNTON, *Managing Editor*

MEREDITH C. WILCOX, *Administrative Director, Illustrations*

STAFF FOR THIS BOOK

LISA THOMAS, *Editor*

SANAA AKKACH, *Art Director*

ADRIAN COAKLEY, *Illustrations Editor*

AL MORROW, *Design Assistant*

LEWIS BASSFORD, *Production Project Manager*

MARSHALL KIKER, *Illustrations Specialist*

MANUFACTURING AND QUALITY MANAGEMENT

CHRISTOPHER A. LIEDEL, *Chief Financial Officer*

PHILLIP L. SCHLOSSER, *Vice President*

CHRIS BROWN, *Technical Director*

NICOLE ELLIOTT, *Manager*

RACHEL FAULISE, *Manager*

The National Geographic Society is one of the world's largest nonprofit scientific and educational organizations. Founded in 1888 to "increase and diffuse geographic knowledge," the Society works to inspire people to care about the planet. It reaches more than 325 million people worldwide each month through its official journal, *National Geographic,* and other magazines; National Geographic Channel; television documentaries; music; radio; films; books; DVDs; maps; exhibitions; school publishing programs; interactive media; and merchandise. National Geographic has funded more than 9,000 scientific research, conservation and exploration projects and supports an education program combating geographic illiteracy.

For more information, please call 1-800-NGS LINE (647-5463) or write to the following address:

National Geographic Society
1145 17th Street N.W.
Washington, D.C. 20036-4688 U.S.A.

Visit us online at www.nationalgeographic.com

For information about special discounts for bulk purchases, please contact National Geographic Books Special Sales: ngspecsales@ngs.org

For rights or permissions inquiries, please contact National Geographic Books Subsidiary Rights: ngbookrights@ngs.org

ISBN: 978-1-4262-0606-1

Printed in U.S.A.

10/CK-CML/1